AESS Interdisciplinary Environmental Studies and Sciences Series

Series Editor

Wil Burns, Forum for Climate Engineering Assessment, School of International Service, American University, Washington, DC, USA

Environmental professionals and scholars need resources that can help them to resolve interdisciplinary issues intrinsic to environmental management, governance, and research. The AESS branded book series draws upon a range of disciplinary fields pertinent to addressing environmental issues, including the physical and biological sciences, social sciences, engineering, economics, sustainability planning, and public policy. The rising importance of the interdisciplinary approach is evident in the growth of interdisciplinary academic environmental programs, such Environmental Studies and Sciences (ES&S), and related 'sustainability studies.'

The growth of interdisciplinary environmental education and professions, however, has yet to be accompanied by the complementary development of a vigorous and relevant interdisciplinary environmental literature. This series addresses this by publishing books and monographs grounded in interdisciplinary approaches to issues. It supports teaching and experiential learning in ES&S and sustainability studies programs, as well as those engaged in professional environmental occupations in both public and private sectors.

The series is designed to foster development of publications with clear and creative integration of the physical and biological sciences with other disciplines in the quest to address serious environmental problems. We will seek to subject submitted manuscripts to rigorous peer review by academics and professionals who share our interdisciplinary perspectives. The series will also be managed by an Editorial board of national and internationally recognized environmental academics and practitioners from a broad array of environmentally relevant disciplines who also embrace an interdisciplinary orientation.

More information about this series at https://link.springer.com/bookseries/13637

Martha Richmond

Cancer Hazards: Parathion, Malathion, Diazinon, Tetrachlorvinphos and Glyphosate

The 2015 IARC Classifications: Implications for Regulation, Environmental Justice, and Global Health

AESS
Association for Environmental Studies and Sciences

Springer

Martha Richmond
Suffolk University
Boston, MA, USA

ISSN 2509-9787 ISSN 2509-9795 (electronic)
AESS Interdisciplinary Environmental Studies and Sciences Series
ISBN 978-3-030-81952-1 ISBN 978-3-030-81953-8 (eBook)
https://doi.org/10.1007/978-3-030-81953-8

This Springer imprint is published by the registered company Springer Nature Switzerland AG
The registered company address is: Gewerbestrasse 11, 6330 Cham, Switzerland

Acknowledgments

I would like to first acknowledge the help of Melinda (Lindy) Paul of Springer-Nature who expressed interest and supported me as I was preparing a proposal for this book. I also want to thank my editors with Springer-Nature for their guidance, and, at times, patience during the time that I was preparing my manuscript. Finally, I would be remiss if I did not thank Walter A. Rosenberg, Former Editor-in-Chief of the Journal of Environmental Sciences and Studies for his interest and encouragement.

To colleagues and friends who gave their time to be interviewed, who provided instructions and suggestions, or who, in other ways, enhanced my knowledge of the material that has gone into the book, I would also like to express my gratitude. I thank Scott Lussier of Suffolk University and Mary Margaret Minnis of Pace University for their excellent instruction in ArcGIS and their suggestions about the ArcMaps that are included in the book. While the maps are only for pesticide and herbicide use in California, given that California is a state with extensive agricultural activity and is also a state maintaining complete annual records on pesticide and herbicide use, they provide an excellent visual representation of how extensively some of the substances discussed in the book are used. To Geoffrey Sunshine, a former colleague who is now associated with Tufts University Medical School, I owe a special thanks. His suggestions and critical comments on the chapter discussing the 2015 IARC evaluations helped me to develop a stronger and more clearly written chapter. Jeannie Economos of the Farmworker Association of Florida provided invaluable information about working conditions for farm, nursery, and landscape workers in Florida. She also identified contacts in California that I might interview. I was fortunate to speak with Anne Katten, Pesticide and Work Safety Project Director of the California Rural Legal Assistance Foundation who provided additional information on environmental justice and advocacy groups in several states. I am particularly grateful that both Jeannie and Anne were available to talk during a time when the COVID pandemic was an issue of great concern, and when time became a highly valued commodity. Finally, I want to thank Emily Bradford Gallup, Information Systems Analyst, for her excellent assistance and suggestions in developing index terms.

As someone with a background in public health, especially environmental health, I have long been interested in the impact of toxins on communities, especially underserved communities. The hazards posed by pesticide use, especially its extensive use in developed and less developed countries, are clearly an area of concern. I was quite fortunate to visit twice and to be sponsored by the University of the Free State in South Africa to visit as a guest lecturer. I was able to meet with several faculty from the QwaQwa campus as well as the main campus in Bloemfontein. Special thanks go to Hennie Claassen of the QwaQwa campus who sponsored my second visit, made the necessary arrangements, and who also spent many hours taking me to visit emerging farms, speak with farmers, and learn more about the issues and challenges regarding farming as well as pesticide and herbicide use in South Africa. Hennie also provided me with opportunities to become better acquainted with the presence of agroindustry in farming communities.

I am fortunate to have had access to the Sawyer Library at Suffolk University, both to locate reference materials in the library collection and also acquire those not available through Suffolk University. This was an especially valuable resource during the COVID pandemic when in-person visits were not possible. I want to thank all the staff members who helped me.

I want to thank my family. Particular thanks go to my immediate family members, including Andrea and Kevin Shurtleff, Jack Shurtleff, Thomas Shurtleff, and Nicole Richmond. Their interest in and excitement about the book and their often-probing questions provided needed guidance for effective framing of issues.

I was very grateful for the constant companionship of our late dog Allegra who sat at my feet during much of the time I spent writing and researching the book manuscript. Her warmth and kindness were critical during these times. And, finally but hardly least, it is impossible to overstate my gratitude for the constant encouragement, interest, and love that my husband, John Berg provided. Writing a book is frequently challenging. Having his enduring support at such times has been important beyond measure. I love him very much and am very fortunate that he is part of my life.

Introduction

From the time that plant cultivation became established, unwanted pests—insects, plants, and molds—have threatened crops, frequently leading to poor yields or even complete losses. To combat pests and minimize loss, farmers throughout history have developed a number of "pesticides." With the exception of a few organic chemicals extracted from natural sources, historically most pesticides were inorganic substances, including some highly toxic compounds such as those containing mercury and arsenic. Many were "home-grown" and used in small amounts rather than produced in large volumes; or they came from natural sources such as pyrethrum, extracted from chrysanthemums. However, with the growth of modern chemical methods in the late nineteenth and twentieth centuries, the development of compounds more targeted to pests and, theoretically, less harmful to humans and farm animals became a possibility. Additionally, these compounds were generally produced by industrial organizations and were more readily available. This development became an incentive for the growth of the large agrochemical industry.

Whatever their etiology, pesticides and herbicides are intended to be acute toxins to the targeted species: to quickly kill unwanted pests or plants, while, in theory, not posing a significant danger to larger populations. This includes the larger ecosystem; farm animals; people working in agricultural, pest removal, or landscape applications; those living in close proximity to such activities; or members of the general public who might consume food or be exposed to fibers from agricultural products treated with the pesticides. Quantitatively, this means that for the target species, acute toxicity, generally measured as LD_{50} or mean lethal dose, is orders of magnitude smaller than the LD_{50} of humans and related species.

A large number of what might be considered the "first generation" of pesticides were chlorinated hydrocarbons. This included substances such as the pesticides dichlorodiphenyltrichloroethane (DDT), dichlorodiphenyldichloroethylene (DDE), dieldrin and endrin, or herbicides such as dichlorophenoxyacetic acid (2,4-D) or trichlorophenoxyacetic acid (2,4,5-T), all of which are deemed to post significant dangers, not only to those working directly with them but to the general population and the larger environment. A distinct disadvantage of these substances is that, because they are lipid-soluble, they bioaccumulate, both in individuals who come into contact with them and also because they concentrate through the food chain. Today, all are deemed to pose significant dangers, not only to those working directly with them but also to the general population and the larger environment.

In contrast, the organophosphate pesticides as well as the herbicide glyphosate, because they are water-soluble, were seen as having a distinct advantage. They were generally metabolized to easily excreted products, or, in some cases, eliminated without metabolic transformations. Thus, while a number of chlorinated hydrocarbon pesticides and herbicides have been banned or highly restricted for use through regulatory processes, especially in developed countries, the use of organophosphate substances has grown dramatically. However, coupled with the development of GM seeds in the 1980s and 1990s, this large growth has led to significant environmental persistence. This was a consideration by the International Agency for Research on Cancer (IARC), in choosing to evaluate four organophosphate pesticides as well as the organophosphate herbicide glyphosate in 2015. Results of their findings were published in a 2017 Monograph entitled, "Some Organophosphate Insecticides and Herbicides." IARC, an agency of the World Health Organization, has published a series of monographs, beginning in 1972. Each publication provides findings and conclusions of expert panels who have examined evidence from peer-reviewed publications and government data to assess the carcinogenic hazards of a variety of materials. The 2017 Monograph details findings from the review panels that concluded tetrachlorvinphos, parathion, malathion, diazinon, and glyphosate pose a carcinogenic hazard.

The reaction to the 2017 Monograph has been strong and frequently polarized. Commercial entities and other organizations allied with or speaking for these groups have challenged the IARC designations in a number of ways: attacking the scientific basis of findings; claiming bias of those participating in the IARC reviews; and stating that the need for food security mandates pesticide and herbicide use. This has been especially true with respect to glyphosate, in part because of the development of Roundup Ready™ genetically modified seed, marketed by Monsanto. Negative responses to the glyphosate classification include publications in the lay press, congressional hearings in the US, extensive scientific publications challenging methods used in IARC assessments, and administrative changes in industries that synthesize and distribute glyphosate-containing products for use in agriculture or landscaping, presumably to lessen potential liability. In contrast, environmental and health advocacy groups have supported the IARC findings: defending the scientific basis of designations, questioning the strength of arguments for food security, and pointing out concerns about environmental spread and persistence.

Often buried in the controversies is the distinction between hazard identification and risk assessment. Hazard identification, the purpose of IARC's review process, derives from a careful examination of vetted scientific data. The goal is to identify substances that can pose a carcinogenic risk under potential exposure conditions. In contrast, risk assessment takes information on hazard identification and applies this information to various models used to develop estimates of risk for various constituencies. The constituencies include the general public, or, within the occupational sector, those who work in agricultural and landscaping industries, and who may require exposure protection. Risk assessment generally incorporates hazard identification into various exposure models to develop quantitative information regarding safe levels of exposure and potential intake.

It is important to note that IARC's hazard assessments are intended to address potential outcomes on a global basis, which in many cases involves conditions where exposures are unknown, or, if known, are not well-regulated. This leads to questions of environmental justice. Those exposed to toxic substances through poor working conditions and lax environmental regulations and who are not adequately protected are far more likely to respond adversely. This can include both short-term reactions and long-term responses, including the development of chronic diseases such as cancer. For various reasons, these responses are often either underreported or not reported at all to the appropriate agencies. Many of those experiencing adverse reactions are members of underserved communities or live and work in underdeveloped/poorly developed geographic constituencies. They are afforded little or no protection and, furthermore, lack strong advocacy groups.

While it can be acknowledged that issues of environmental justice are often specific to given agricultural sites or specific agricultural operations, food and flower production have also become increasingly global. It is unclear how, in the production-export process, agribusiness is regulated. Hence, the potential for adverse effects of pesticide and herbicide use becomes an issue of growing complexity.

The following book focuses on the four pesticides and the herbicide assessed by IARC in 2015. It first provides a historical perspective for the introduction of each compound into agriculture and landscaping uses, documents the spread in use and the environmental impact of each, and critiques data on health effects. Particular focus will be given to chronic diseases, especially cancer or pre-cancerous conditions. The second part of the book looks specifically at the IARC review process, both in terms of its development, its general approach, and then on specific issues of the 2017 IARC Monograph. The next chapters detail and critically examine the responses to the IARC designations and implications of these responses. The final sections will discuss the larger issues of environmental justice, both in specific environments or industries as well as the global community.

Contents

Part I
Overview of Parathion, Malathion, Diazinon, Tetrachlorvinphos and Glyphosate

Chapter 1
Discovery and Commercial Introduction and Mode of Action of Parathion, Malathion, Diazinon, Tetrachlorvinphos, and Glyphosate

Abstract The development of the organophosphate pesticides considered by the International Agency for Research on Cancer (IARC) originally came from the synthesis of more potent neurotoxins during World War II when it was discovered that some synthesized neurotoxins were effective against insects, but far less lethal for higher animals than the more potent neurotoxins. In contrast, glyphosate, while having some structural similarities to the organophosphates, is not an insecticide, but is effective against the growth of unwanted plants, with a low level of acute toxicity to animals. The following chapter provides a background for the development of each of the five substances considered by IARC, discusses their mode(s) of action, and summarizes their uses and overall potency, including "safety factors" of human metabolism.

1.1 Discovery and Commercial Introduction of Organic Phosphate Pesticides: Parathion, Malathion, Diazinon, and Tetrachlorvinphos

1.1.1 Discovery and Commercial Development of Organophosphate Pesticides

The first of the pesticides to be developed, parathion, was synthesized in 1944 by Gerhard Schrader, a chemist working for IG Farben, the chemical conglomerate formed through the merger of BASF, Bayer, Hoechst, Agfa, Chemische Fabrik Griesheim-Elektron, and Chemische Fabrik Griesheim vorm. Although Schrader was involved in the synthesis of highly toxic organophosphate compounds such as tabun and sarin, he was purportedly more interested in developing organophosphates that could be used to control agricultural pests, allowing for production of more crops to further Germany's war effort. In the early 1940s, he synthesized parathion and identified it as a potential insecticide because of its lethality for insects. While,adjusted for body weight, the difference between the LD_{50} (lethal dose for half of a test population

M. Richmond, *Cancer Hazards: Parathion, Malathion, Diazinon, Tetrachlorvinphos and Glyphosate*, AESS Interdisciplinary Environmental Studies and Sciences Series,
https://doi.org/10.1007/978-3-030-81953-8_1

assuming a normal distribution of lethality) of parathion for insects, 0.7–1.0 μg/g for honeybee; or 0.5 μg/g for a housefly [8] versus 2 mg/kg for humans [5] is not appreciably different, far less parathion was seen as necessary to combat insect pests. Parathion is also effective against insects in all developmental stages.

Following World War II, IG Farben as a chemical conglomerate was dismantled. The immediate intent was to disassemble the German chemical companies, although, with time and in response to the cold war, in Western Germany several of the original member companies of IG Farben were re-established. This included Bayer in 1951; Hoechst in 1951, and BASF, re-established in 1953. Hoechst ultimately became a holding company for many other small chemical companies that had been disassembled after the war.

In the postwar period, many records of organophosphate development were lost or destroyed. Thus, while it was possible to establish that parathion was developed by Schrader in 1944, data on the development of malathion, which he may also have synthesized were not easy to obtain. However, western industries, especially those in the US, became interested in both chemicals for pest control, and they were further developed for commercial use. Parathion was introduced in the US in 1948. Malathion has been used in the US since 1950.

The remaining two organophosphate pesticides were synthesized following World War II. Diazinon was developed by Ciba-Geigy in 1952 and has been commercially available for use since the mid-1950s. Tetrachlorvinphos was registered for use in 1966. The initial registrant was Shell Chemical Company, which in turn transferred the registration to E.I. DuPont de Nemours. In September 1992, DuPont transferred the registration of technical tetrachlorvinphos to Boehringer Ingelheim Animal Health Inc. and Hartz Mountain Corporation [2].

1.1.2 Basic Principle of Organic Phosphate Toxicity

Although they share some structural characteristics, the four organic phosphate pesticides as a group have slightly different modes of toxicity. The structural formulas of the four compounds are shown in Fig. 1.1.

While the structures of the four pesticides differ in a number of aspects, they all share common motifs that facilitate their binding to acetylcholine esterase, an enzyme necessary for cholinergic nerve transmission. In cholinergic neurotransmission, acetylcholine is released by the axon of a nerve fiber and binds to a receptor site either on an adjacent dendrite or on a muscle cell. Regardless of whether the receptor cell is another neuron or a muscle cell, the acetylcholine must be broken down by an acetylcholine esterase before another neurotransmission event can take place. If the acetylcholine esterase is inhibited, no additional transmission can take place. The process of neurotransmission is shown in Fig. 1.2.

The presynaptic nerve terminal releases acetylcholine into the synaptic cleft. Acetylcholine binds to receptors on the postsynaptic nerve terminal. A choline

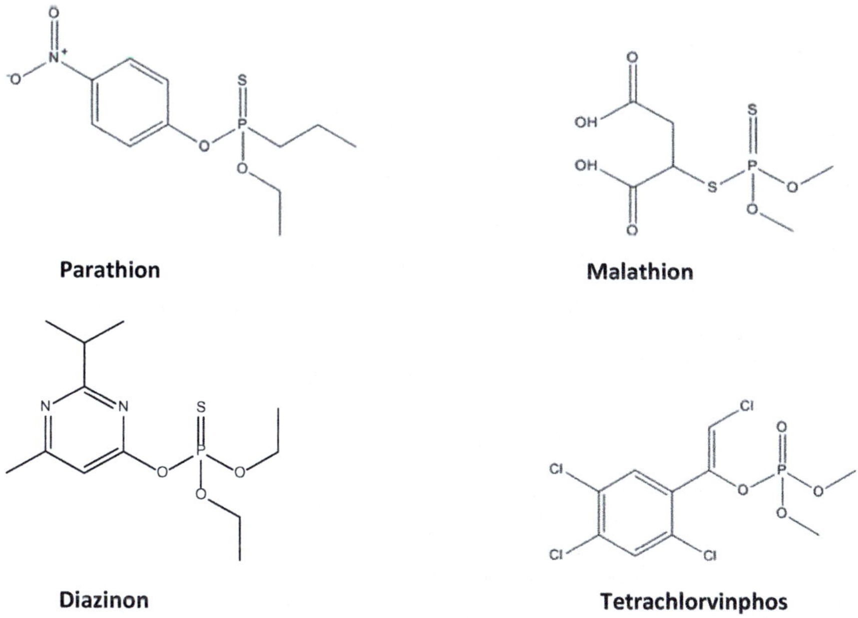

Fig. 1.1 Structures of parathion, malathion, diazinon, and tetrachlorvinphos

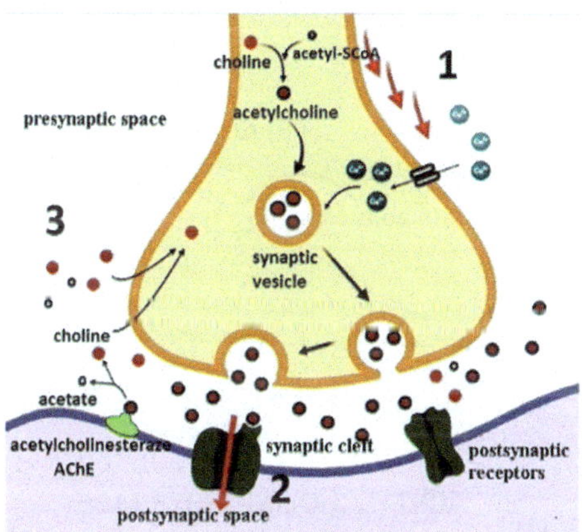

Fig. 1.2 Neurotransmission involving a cholinergic synapse

esterase must hydrolyze the bound acetylcholine before another neurotransmission can occur.[1]

[1] Permission to use figure provided by Gabriela Dumitrita Stanciu, first author of ***Alzheimer's Disease Pharmacotherapy*** in ***Relation*** to ***Cholinergic System Involvement***. by Gabriela Dumitrita Stanciu. Andrei Luca. … ***Biomolecules 2020, 10(1), 40***; https://doi.org/10.3390/biom10010040.

The normal process of acetylcholine breakdown by acetylcholine esterase is shown in Fig. 1.3. For comparison, Fig. 1.4 illustrates inhibition by an organophosphate compound. In the form shown, the organophosphate can be hydrolyzed from the bond it has formed with serine, and thus the inhibition is considered reversible. However, if an R group attached to phosphate is removed by hydrolysis, the phosphate remains bound to serine and the enzyme inhibition is irreversible. This process is known as aging. The aging process is illustrated in Fig. 1.5.

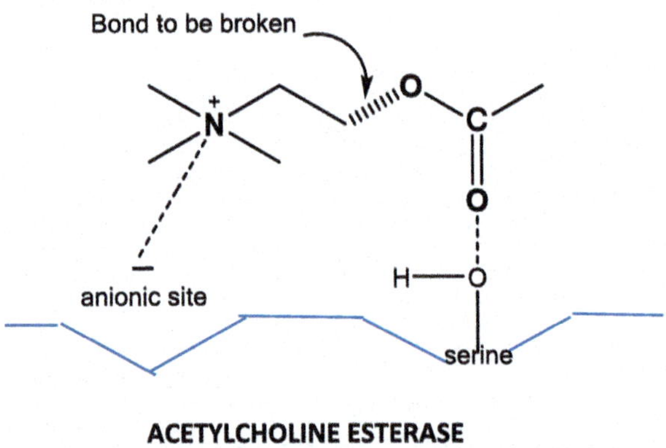

Fig. 1.3 Normal process of acetylcholine breakdown by acetylcholine esterase. The neurotransmitter is bound to the enzyme in two places. The binding facilitates breakdown by the esterase

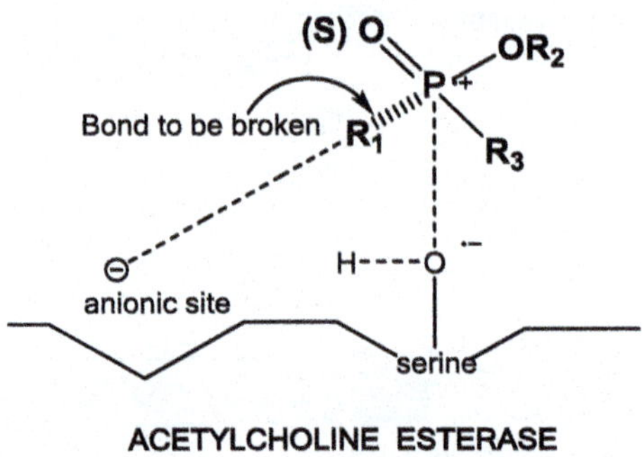

Fig. 1.4 Inhibition of acetylcholine breakdown by organophosphates. The phosphate group binds to the serine site of the enzyme. Until the bond between R_1 and phosphorus is broken, no new nerve transmission can occur. However, ultimately the bond is broken and neurotransmission is restored

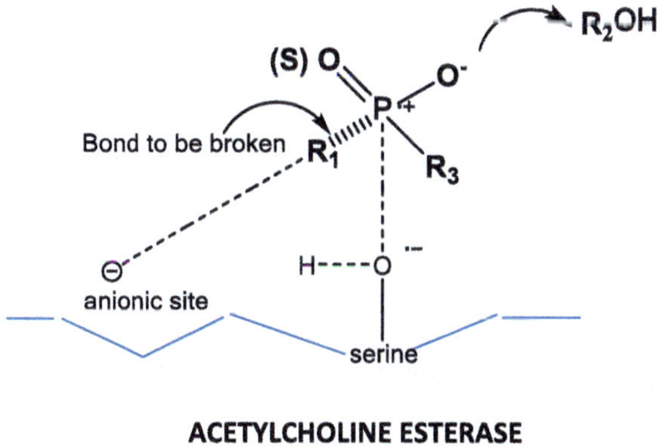

Fig. 1.5 Aging of organophosphate. If the R_2 group attached to phosphate is removed, the attractive forces holding the phosphodiester group are sufficiently strong to hold the complex onto the acetylcholine esterase, rendering it inactive

1.2 Discovery and Commercial Introduction of Glyphosate

1.2.1 Discovery and Commercial Development of Glyphosate

Glyphosate was synthesized in 1950 by Henri Martin, a Swiss chemist, who was conducting research on the development of new pharmaceutical agents. Although it was not found to have pharmaceutical properties, glyphosate was patented because of its properties as a chelating or metal-binding compound. The observation that glyphosate binds to manganese, a metal involved in the shikimic acid pathway necessary for the formation of phenylalanine, tyrosine, and tryptophan, amino acids found in all proteins which plants, fungi, and bacteria must synthesize later led to glyphosate's development as an herbicide capable of killing unwanted plants, while theoretically, not harming animals. As noted by Dill et al. [1] glyphosate also was shown to have a very low acute toxicity, with an LD_{50} said to be comparable to sodium chloride. By comparison to a number of commercial herbicides available in the early 1970s, taken together with its water solubility (rendering it easily excreted) it was seen as having ideal potential as an agent for killing unwanted plants. Glyphosate was marketed by Monsanto as Roundup™ in 1974. Monsanto held the patent until 2000.

Although Monsanto's patent expired in 2000 and glyphosate is now marketed by a number of other companies, Monsanto was able to continue marketing Roundup™ through the development of Roundup™ Ready crop seed. This was done by identifying a glyphosate-resistant strain of bacteria, *Agrobacterium* sp. strain CP4, isolating and cloning the gene conferring resistance, and inserting it into the plant genome. "Roundup™-Ready" seed stock for a number of crop seeds has been marketed to commercial farming operations and is generally sold together with Roundup™ [3].

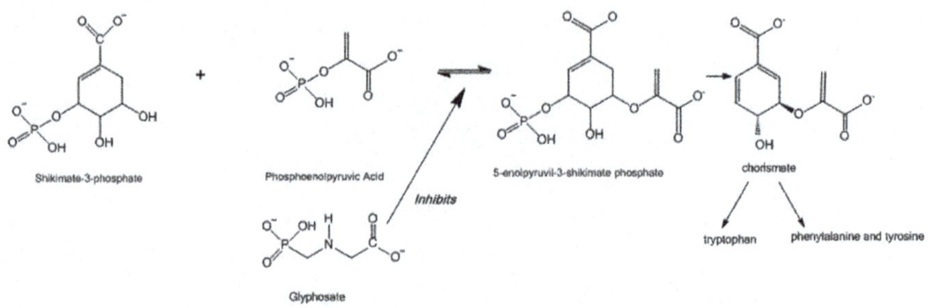

Fig. 1.6 The biosynthetic pathway leading to 5-enoyl-3-shikimate phosphate with an outline of further steps in phenylalanine, tyrosine, and tryptophan synthesis

1.2.2 *Basic Principle of Glyphosate Toxicity*

As noted, glyphosate inhibits the shikimic acid pathway, a pathway found in plants, bacteria, and fungi, but not in animals. Hence, it stops plant growth. In contrast, animals, because they are unable to synthesize the three amino acids, must take them in from dietary sources. The biosynthetic pathway for the three amino acids and the point of glyphosate inhibition are outlined in Fig. 1.6.

As noted earlier, glyphosate is a chelating agent, a compound that binds to metal ions. It was originally felt that glyphosate's inhibitory effect is, in part, because of its ability to bind to manganese, an ion needed for the synthesis of 5-enoyl-3-shikimate phosphate [4, 6]. More recent findings, however, indicate that it is quite likely that the inhibitory effect primarily results from glyphosate binding to the enzyme necessary for 5-enoyl-3-shikimate phosphate synthesis because it competes with phosphoenol-pyruvic acid for the enzyme binding site [7].

1.3 Purpose of Use and Toxicity for Target Species and Humans

While all four organophosphates are pesticides and all are neurotoxins, binding to acetylcholine esterase, their present purpose for use as well as acute toxicity to target species and humans vary. Glyphosate acute toxicity is seen as limited to plants and microorganisms that employ the shikimate pathway in synthesizing phenylalanine, tyrosine, and tryptophan. Table 1.1 provides information on each substance, including the year of introduction, human acute toxicity, trade names under which each substance is marketed, present use, and target species for each.

Table 1.1 Summary data for parathion, malathion, diazinon, tetrachlorvinphos, and glyphosate

Agent	Year introduced	Human toxicity	Trade names	Present usw
Parathion	1948	(estimated values) Minimum LD 0.17 to 1.47 mg/kg; estimated LD_{50} between 5 and 50 mg/kg[1]	Alkron; Alleron; Bladan; Bladan F; Corothion; Ethlon; Folidol; Fosfermo; Orthophos; Panthion; Paradust; Paraphos; Thiophos	Present use canceled in many countries; when used is for control of insects and mites on food crops
Malathion	Used in the US since 1950	Revised IDLH: 250 mg/m[1,2]	At least 17 different trade names, including Agrothion, Heckthion, Hilmala, Hilthion, Malatox, and Tragumal[3]	Broad-spectrum insecticide used commercially for food, and feed, to store grain, on lawns food and feed crops, grain facilities, lawns, gardens, nurseries roads, cattle ranges, mosquito abatement, and head and body lice
Diazinon	1952	The acute reference dose (RfD) is 0.0025 mg/kg/day. The chronic RfD is 0.0002 mg/kg/day[2,3]	Diazinon is sold as diazinon as well as Alfatox, Basudin, AG 500, Dazzel, Gardentox, Diazide, Spectracide and Knoxout	Commercially used to control soil and foliage insects and pests on fruit, vegetable, nut, and field crops Used as an insecticidal ear tag for non-lactating cattle
Tetrachlorvinphos	1966	LD_{50} = 1480 mg/kg (M) 465–965 mg/kg (F) Oral dose; LD_{50} > 2000 mg/kg Dermal dose; LC_{50} > 3.61 mg/L Inhalation[1]	Appex, Dust M, Gardcide, Gardona, Rabon, Rabon Oral Larvicide (ROL), and Rabond Ravap Stirophos, Stirofos,Stitofos, Debantic, Dietreen, SD, Shell SD	Used to control fleas, ticks, various flies, lice, and insect larvae on livestock and pests. Can be distributed in the perimeter of livestock. It can be used to control fruit pests, vegetable pests, cereal, cotton, corn, rice, and tobacco pests

(continued)

Table 1.1 (continued)

Agent	Year introduced	Human toxicity	Trade names	Present usw
Glyphosate	1974	Human acute toxicity data not readily available; animal acute toxicity depressed body weight or death < 14-day exposure to 3000–3500 mg/kg/day[1,2]	Roundup™ Watkins Weedkiller McGregor's Weedout Agpro Glyphosate 360 Dow Glyphosate 360 Glyphosate 360 Gro-Chem Glyphosate 360 Hortcare Glyphosate 360 SC Jolyn G 360® Nufarm Glyphosate Gold Samurai TM Orion Glyphosate 360 Proactive Glyphosate 360 Proactive Glyphosate Green Rainbow & Brown Glyphosate	Used for weed control in a variety of circumstances: landscaping; removal along train tracks, powerlines, pipelines; removal of weeds after crop harvesting, and between rows in orchards. Used in crop spraying of a variety of Roundup Ready™ farm crops

[1] {ATSDR, 2019 #68} National Institute for Occupational Safety and Health Malathion CDC https://www.cdc.gov/niosh/idlh/121755.html
[2] IARC Monograph, 2017
[3] Reference dose, RfD is defined as the amount an individual could be exposed to (acute and chronic exposure) for the rest of his/her life without adverse effects.

References

1. Dill GM, Sammons R, Feng PCC, Kohn F, Kretzmer K, Mehrsheikh A, Bleeke M, Honegger JL, Farmer D, Wright D, Haupfear EA (2010) Glyphosate: discovery, development, applications and properties. In: Nandula VK (Ed) Glyphosate resistance in crops and weeds: history, development, and management. John Wiley and Sons, Inc., Hoboken, 1–33. doi:https://doi.org/10.1002/9780470634394
2. Office of Pesticide Programs (2006) Reregistration Eligibility Decision for Tetrachlorvinphos (TCVP). Washington, DC
3. Funke T, Han H, Healy-Fried ML, Fischer M, Schonbrunn E (2006) Molecular basis for the herbicide resistance of Roundup Ready crops. Proc Natl Acad Sci US 103:13010–13015. https://doi.org/10.1073/pnas.0603638103
4. Mertens M, Höss S, Neumann G, Afzal J, Reichenbecher W (2018) Glyphosate, a chelating agent—relevant for ecological risk assessment? Environ Sci Pollut Res 25:5298–5317. DOI: https://doi.org/10.1007/s11356-017-1080-1
5. Reuber M, Susan C (2007) Beyond pesticides/NCAMP factsheet: parathion (ethyl parathion). Chem Watch Fact Sheet. https://www.beyondpesticides.org

6. Samsel A, Seneff S (2015) Glyphosate, pathways to modern diseases III: manganese, neurological diseases, and associated pathologies. Surg Neurol Int Mar 24, 6:45. doi: https://doi.org/10.4103/2152-7806.153876
7. Schonbrunn E, Eschenburg S, Shuttleworth WA, Schloss NA, Evans JNS, Kabsch W (2001) Interaction of the herbicide glyphosate with its target enzyme 5-enolpyruvylshikimate 3-phosphate synthase in atomic detail. Proc Natl Acad Sci US 98(4):1376–1380. DOI: https://doi.org/10.1073/pnas.98.4.1376
8. Stenersen J (2004) Chemical pesticides mode of action and toxicology. CRC Press, Boca Raton, p 103

Chapter 2
Use: Growth in Use, Present-Day Volume, and Projected Use of Parathion, Malathion, Diazinon, Tetrachlorvinphos, and Glyphosate

Abstract To understand the environmental spread of both organophosphate pesticides and glyphosate, it is important to consider how the use of many has grown. While not universally true, since initially marketed, the volume of use for most has grown dramatically. At the same time, understanding the extent of use is challenging. Changes in product patents, proprietary trade information, regulatory restrictions, and other factors make it difficult to fully estimate how much is used, especially on a global basis. The following chapter examines these issues. It also considers the most currently available data for the state of California, which, in the US, has the most detailed information.

2.1 Growth in Use, Present-Day Volume, and Projected Use of Parathion, Malathion, Diazinon, Tetrachlorvinphos, and Glyphosate

2.1.1 Parathion

Although initially thought to be an appropriate agent for pest control by Gerhard Schrader, its developer, and by the companies that marketed it in the late 1940s, parathion's very high human acute toxicity has, over time, led to a significant reduction in marketing and use. According to an article published in the New York Times (NY Times) in 1991, parathion did not become a "popular" pesticide until DDT was banned by EPA in 1972 [17]. As more and more environmental experts and advocates expressed concern, industry arguments, including those of large-scale farmers, focused on the lower price of parathion as compared with other organophosphate pesticides [17]. The NY Times article noted that in 1991, EPA scientists recommended that parathion sales in the US should be banned, although a final EPA decision was not made, and despite the EPA scientists' recommendation in 1991, it was not until 2000 that EPA posted a decision to cancel re-registration of parathion under its various commercial names. In its notice of cancellation, EPA noted that:

M. Richmond, *Cancer Hazards: Parathion, Malathion, Diazinon, Tetrachlorvinphos and Glyphosate*, AESS Interdisciplinary Environmental Studies and Sciences Series,
https://doi.org/10.1007/978-3-030-81953-8_2

> Due to the high estimated risks based on the best information available to the Agency, and the registrants' decision not to support the data requirements for reregistration, the registrants have signed an agreement to voluntarily cancel their registrations. Cancellation of manufacturing use products is effective immediately, and end use products will be canceled effective December 31, 2002, with last legal use on October 31, 2003. Additionally, ethyl parathion end use product labels will be amended to delete use on corn grown for seed, the site with the highest potential risk to reentry workers, and add the last legal use date [12].

Globally parathion is now banned or its use is severely curtailed in a large majority of countries. In spite of this, however, periodic reports of poisoning (intentional or non-intentional) from a number of countries emerge, suggesting that it may still be used. In the US, the last available data for parathion use comes from California in 2017 where a total of 2.81 lb were used in Alameda County [8]. According to the Agency for Toxic Substances and Disease Registry (ATSDR), across the US small amounts of parathion were used from 2002 to 2014, ranging from a high value of 1,542 lb in 2006 to a low value of 22 lb in 2014 [3]. It was noted, however, that according to the CalEPA, "[the] pesticide might actually have been another pesticide, but misreported each year as parathion." The US Geological Survey, which conducts a nation-wide survey of agricultural pesticide use in the US and territories annually, reported a total annual use of 131.3 lb in 2012 [5], the last date when information was available about agricultural use.

As is true with all pesticides and herbicides, the specific agent may be marketed under a variety of different names, and it is challenging to locate current accurate data on marketing and sales on a global basis. According to the Monograph 112 the production of parathion exceeded 1000 tons/year in 2004 and was reported in 2015 to be produced in China, El Salvador, Germany, and the US [14]. Although spraying on fruit or food crops is banned in China, a publication examining an analytic technique recently developed to detect parathion and differentiate it from malathion found residues on carrots and cucumbers bought at a local market [15]. Parathion in food crops has also been reported in Ghana [10].

2.1.2 Malathion

Malathion is also said to have been developed during World War II, was introduced in the US in 1950, and continues to be used for a variety of agricultural and landscaping activities as well as for mosquito abatement and treatment of head and body lice.

As is true of parathion, malathion use grew following the removal of DDT and related organochlorine pesticides from the market. The ATSDR reports that in 1972 roughly 12,000 US tons were produced [1], and by 1978 the use had grown to 15,500 US tons or 14,000 metric tons [14]. Global use is thought to have reached its maximum value in 1999 when it was widely used in the US to combat the cotton boll weevil. The last data available from the US Geological Survey reported that a total of 913,394 lb or 457 US tons were used agriculturally in the US in 2012 [5].

California state-wide data for 2017 found that 324,282.45 lb or roughly 162 tons of malathion were used in farming, landscaping, and other operations [8].

It is difficult to find good information about the global production of malathion in 2019–2020, and from that to develop a more accurate value for global use. In 2015, as reported by IARC, malathion was manufactured by 49 producers in 10 countries. The largest number of companies were located in China, with other major suppliers found in India, Singapore, the US, the United Kingdom (UK), Denmark, Egypt, Japan, Mexico, and Switzerland [14]. While the global distribution of suppliers may have changed somewhat in the last five years, it is clear that a number of global markets and suppliers for malathion exist.

A 2019 Washington Post article [9] noted that Rachel Carson expressed concern about malathion in "Silent Spring," noting that "the alleged 'safety' of malathion rests on precarious ground," but "as often happens—this was not discovered until the chemical had been in use for several years." Despite this early concern, malathion continues to be used widely and with little regulation. The EPA, while initially planning a re-regulation evaluation, recently canceled this activity. In contrast, UTZ-certified, a Dutch organization that is part of the Rainforest Alliance reported in 2015 that malathion was placed on a pesticide watch list because of its reported toxicity to bees. This means that malathion cannot be used in the production of any UTZ-certified crops. In this program substances can be used on a certified crop only if the following criteria are met:

1. All steps of integrated pest management (IPM) have been followed.
2. Less hazardous alternative pesticides are not available.
3. Specific recommendations are followed to mitigate or reduce the risks related to the hazardous nature of the product [18].

A number of other geographic districts, including some EU countries, Canada, and smaller geographic jurisdictions have restricted uses and exposure to malathion, but it is, to date, not a banned chemical in any large-scale way.

2.1.3 *Diazinon*

Diazinon was initially used commercially to control insects and other pests in fruit, vegetable, and other agricultural products. It has also been used to control mosquitos and other biting insects and was used at one time on golf courses. Another early application was for the control of termites, cockroaches, and flying ants. Some of this use involved residential pest treatment (by licensed applicators or private users). However, in 2004 EPA, citing information on ecosystems and damage to waterfowl, canceled registration for all residential use [2].

Monograph 112, quoting Farm Chemicals International, notes that diazinon is produced in 11 countries. Reported data indicated that commercial production involved 22 companies in China, six in India, five in the US, four in Singapore, three in the UK, and one each in Canada, Israel, Japan, Mexico, Taiwan, and Thailand [14]. However, it is difficult to get accurate information regarding the amount of production by each company or the volume of use, either in specific countries or globally, although chronological data suggest that the use is decreasing. Information from the US indicates that in 1990, 4.67 million kg (4670 metric tons or 5147 US tons) of diazinon were produced [2]. Annually, in that same year, the use was 5900 metric tons or 6503 US tons [14]. When all residential uses of diazinon (accounting for 70% of all use) were canceled, the annual consumption decreased significantly. Data available from US Geological Survey indicate that in 2012, the last year for which US Geological Survey data are available, US commercial application was 79,562.2 kg or 88 US tons [5] State-wide California data for 2017 reported a total of 40,266.21 lb or approximately 20 US tons were used.

The use of diazinon has been either canceled or restricted in a number of countries or geographic unions. This includes the US and Canada, both of which have greatly restricted their applications. The EU has withdrawn approval of diazinon for the production of crops and for the treatment of farm animals, and in France permission has been withdrawn for diazinon in flea collars for domestic animals [14]. Australia has placed restrictions on diazinon for lice control, including control among domestic and farm animals. In 2007, the Australian Pesticides and Veterinary Medicines Authority banned diazinon for lice control in sheep [4], although it is used for other purposes. Thus, given that diazinon production continues on a global basis, it is likely that a significant amount is being employed in those countries that have not severely restricted or banned its use.

2.1.4 *Tetrachlorvinphos*

Tetrachlorvinphos, introduced in 1966, is employed primarily in lice, flea, and tick control, or to remove larvae on domestic and farm animals in commercial applications. It has also been used to control pests of fruit, vegetable, corn, cotton, cereal, rice, and tobacco. Residential applications include sprays and powders to control fleas and ticks on domestic animals, as well as flea and tick collars.

In 2020, as compiled by PubChem, a total of six vendors produced tetrachlorvinphos. Five of these are in China, and the remaining one is in Canada [6].

In 1987, the US EPA canceled all voluntary crop uses for tetrachlorvinphos, and in 2008 EPA began the re-registration review process. As part of that process, the National Resources Defense Council (NRDC) submitted a petition to EPA asking that, based on a finding published in 2008 [11] regarding human health effects, all tetrachlorvinphos uses be canceled for domestic animals. As of 2017, the decision to cancel or restrict registration is pending. In large part, this is because EPA asked vendors to provide additional information addressing several human health issues of concern [13]. While IARC reported that approximately 400 metric tons (approximately 363 US tons) were used in the US in 2002 [14], the US Geological Survey data reported in 2015 did not include any data regarding the volume of agricultural use of tetrachlorvinphos [5]. California state-wide data for tetrachlorvinphos found that a total of 23.30 lb or 0.012 US tons were used in agriculture and landscaping [8].

In 2003, the EU revoked approval for tetrachlorvinphos. It was used for slightly different periods in France and The Netherlands, although it is not used to any appreciable extent in 2020. According to IARC tetrachlorvinphos use was registered in Canada, South Africa, and Australia in 2015 [14].

2.1.5 Glyphosate

As noted earlier, glyphosate, originally patented as a metal-binding or chelating agent, was found to inhibit a biosynthetic pathway unique to plants, fungi, and microorganisms needed for the synthesis of three amino acids. Because animals must take these three amino acids in through their diet, and because of its low acute toxicity, glyphosate was seen as an ideal plant control agent. Glyphosate can be prepared as several different salts, a property that enhances its water solubility. Initially introduced and marketed by the Monsanto Chemical Company in 1974 as Roundup™, the Monsanto patent expired in 1991 outside the US and in 2000 in the US [14]. It is now marketed under several different names in a number of countries.

As an herbicide, glyphosate was initially used to control weeds in landscaping; along train tracks, power lines, and pipelines; to remove weeds after crop harvesting, and between rows in orchards. However, following Monsanto's introduction of Roundup Ready™ seed in the 1990s, glyphosate use increased significantly. The original Roundup Ready™ seeds were those for several large-scale farm crops, including soybean, corn, and cotton. In subsequent years, a number of other Roundup Ready™ seeds have been introduced, including several other important commercial crops: canola, sugarcane, and sugar beets as well as crops grown on a smaller scale. Glyphosate is also used residentially and has been employed as an aerial spray to destroy cocaine and marijuana production in Latin America [14].

The introduction of Roundup Ready™ seed has resulted in a dramatic growth in the use of glyphosate. This is primarily because a combination of the seed and the pesticide use results in an ability to control weeds and to employ no-till farming methods during plant growth. Today, glyphosate consumption is large in volume and is global. As reported by IARC [14] in 1987, prior to the introduction of genetically modified (GM) seed, approximately 4,000 metric tons (roughly 3,600 US tons) were used in the US. This grew to approximately 80,000 metric tons (roughly 73,000 US tons) in 2007. Data on US agricultural applications available from the US Geological Survey indicate that, in 2012, 130,000 metric tons (118,000 US tons) were used in agriculture alone [5]. State-wide California data reported in 2017 a total of 11,739,800 lb or approximately 5,870 US tons of glyphosate were employed [8]. It is important to point out that these estimates, while based on agricultural reports, must consider the specific chemical salt of glyphosate applied, and because their weights differ somewhat, this introduces a small degree of uncertainty in the actual values calculated.

In 2016, Charles Benbrook [7] published the results of a study examining both the use and growth of glyphosate use in the US and globally. While the study drew from data reported by the US Geological Survey as well as EPA,[1] it primarily used data published by the US Department of Agriculture National Agricultural Statistics Service, which collects statistics on a variety of crops, compiling information for fruit crops and field crops bi-annually in alternating years. Among Benbrook's findings was the observation that glyphosate use from 1974 to 2014 increased 200-fold, and that agricultural applications contributed 90% to this increase.

Global use of glyphosate has also grown considerably. This has been driven in part by the introduction of Roundup Ready™ seed, but also, to a lesser extent, by the number of glyphosate products marketed after the Monsanto patents expired in the US and internationally. According to Benbrook [7], agricultural use of glyphosate increased 14.6-fold from 1996 (when GM seed became available) until 2014. Globally, non-agricultural use increased five-fold.

Global patterns vary. In 2016, the US, Argentina, and Brazil were reported to be the largest consumers of glyphosate on a global basis, but other areas also employed large amounts. In Asia, China and India were the primary users. In Africa, South Africa (as reported) was the greatest employer. Interestingly, consumption by EU countries is variable [7]. As summarized in 2018 [16] while glyphosate applications increased dramatically, the variability from country to country is in part not only because of regulations but also economic issues such as the cost of Roundup Ready™ seed. Overall glyphosate applications have been in large-scale agricultural operations, with the major products being soy, cotton, and maize (corn).

While information about the volume of production of glyphosate is proprietary, the available information in 2020 indicates that China is the major producer, with a total of 13 companies listed as marketing it (Market Watch, 2020). Bayer and Bayer CropScience, the only other major producer, is located in Germany. It is noteworthy that Bayer and Bayer CropScience are "spin-off" companies from the original producer, Monsanto, as a result of corporate re-organization.

Maps of the use of malathion, diazinon, and glyphosate in each California county in 2017 are provided in Figs. 2.1, 2.2, and 2.3. While these maps cannot provide a country-wide visual representation of use, given that California is one of the largest producers of a variety of crops, they do provide information about the extent of use.

[1] EPA data which comes, in part, from registrants, includes estimates for non-agricultural use.

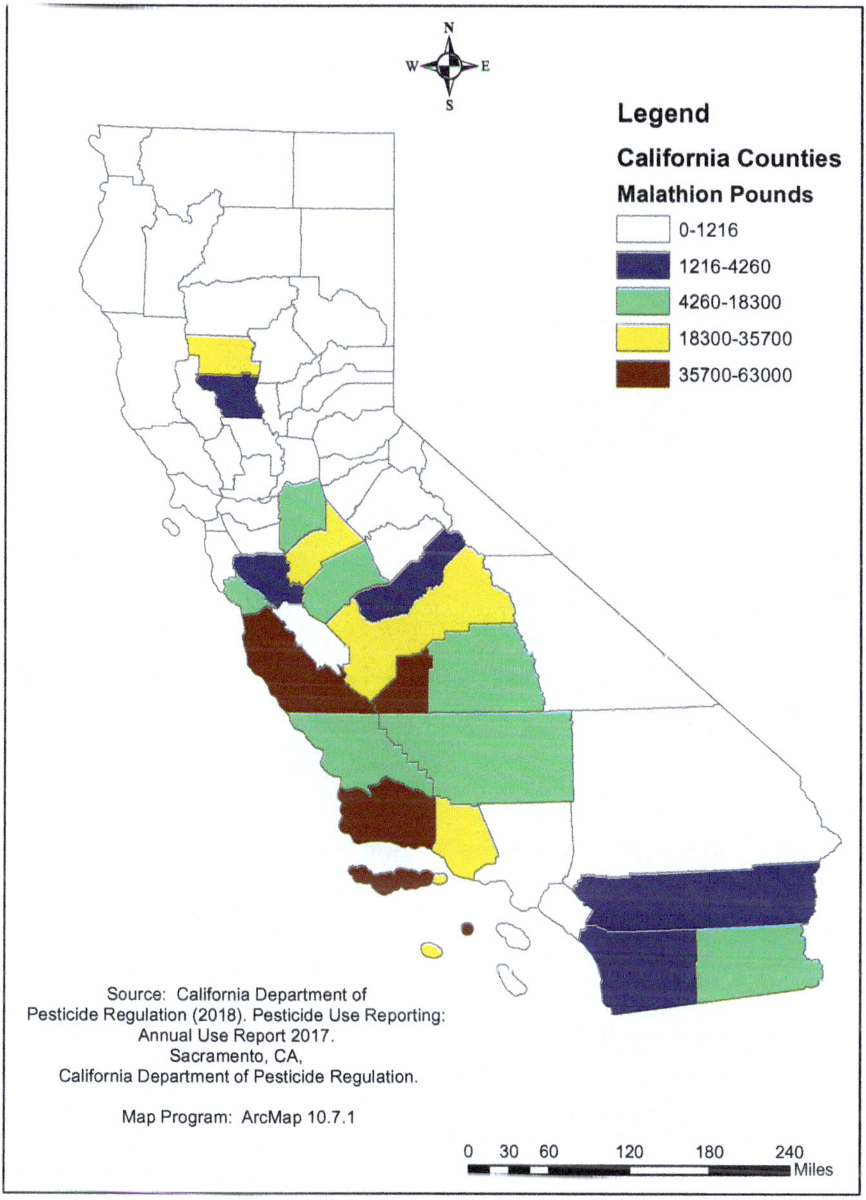

Fig. 2.1 Malathion use in California Counties

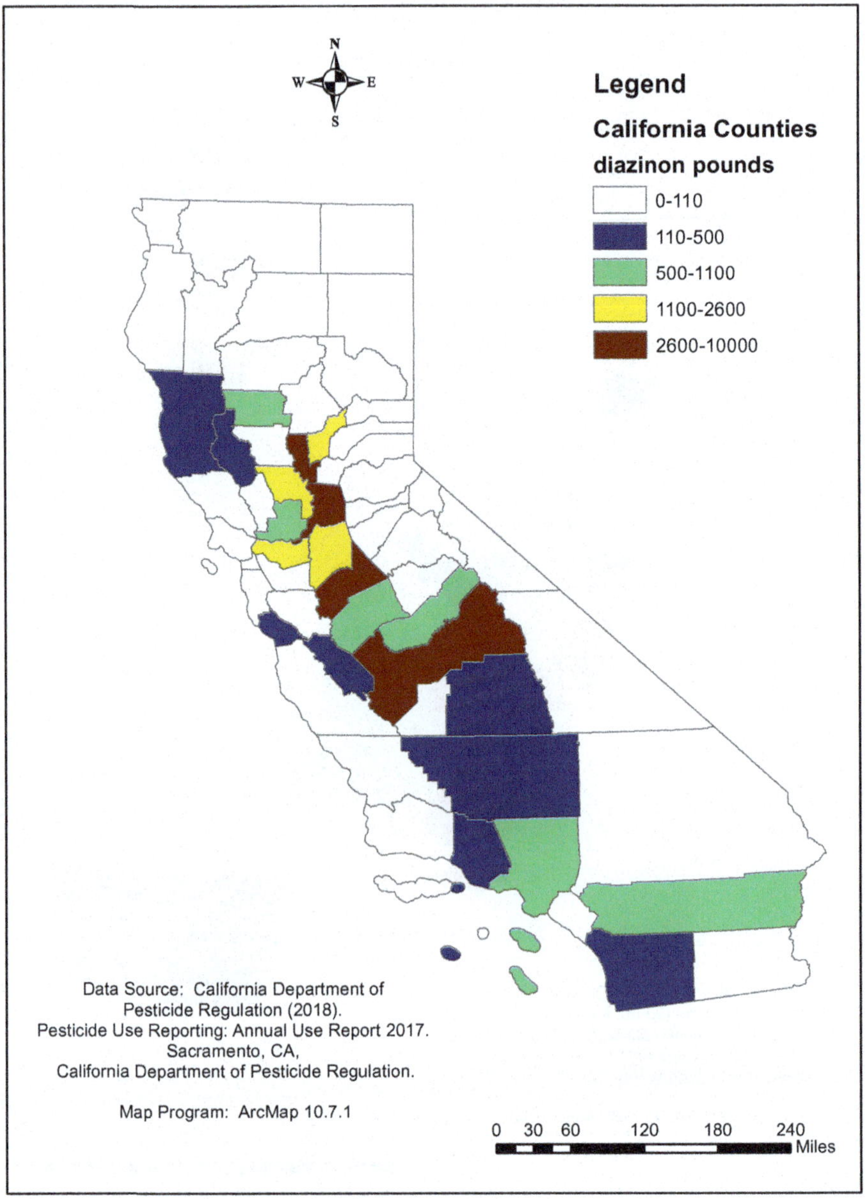

Fig. 2.2 Diazinon use in California Counties

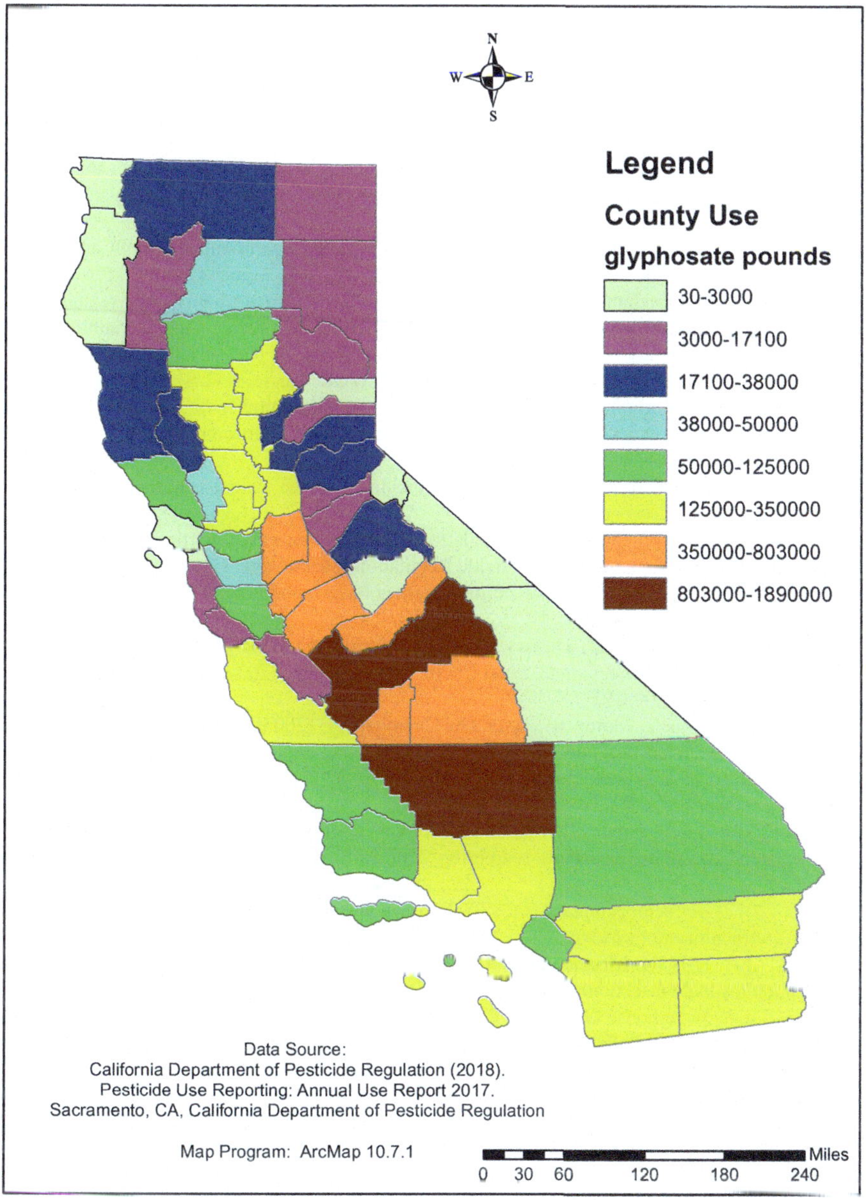

Fig. 2.3 Glyphosate use in California Counties

References

1. ATSDR (2003) Toxicological profile for malathion. US Department of Health and Human Services, Public Health Service. Agency for Toxic Substances and Disease Registry. Atlanta, GA.
2. ATSDR (2008) Toxicological profile for diazinon. US Department of Health and Human Services, Public Health Service. Agency for Toxic Substances and Disease Registry, Atlanta, GA.
3. ATSDR (2019) Toxicological profile for parathion. US Department of Health and Human Services, Agency for Toxic Substances and Disease Registry, Atlanta GA. p. 177
4. Australian Wool Exchange (2019). Alternatives to diazinon. Australian Wool Exchange Ltd
5. Baker NT, Stone W (2015) Estimated annual agricultural pesticide use for counties of the conterminous United States, 2008–12: U.S. Geological survey data series, pp 907–916
6. PubChem [Internet]. Bethesda (MD): National Library of Medicine (US), National Center for Biotechnology Information; 2004-. PubChem Compound Summary for CID 5284462, Tetrachlorvinphos; [cited 2021 Sept. 26]. Available from: https://pubchem.ncbi.nlm.nih.gov/compound/Tetrachlorvinphos
7. Benbrook C (2016) Trends in glyphosate herbicide use in the United States and globally. Env Sci Eur 28(3)
8. California Department of Pesticide Regulation (2018) Pesticide use reporting: annual use report 2017. California Department of Pesticide Regulation, Sacramento, CA
9. Conis E (2019) "Why both major political parties have failed to curb dangerous pesticides." Washington Post. Washington, DC
10. Darko G, Akoto O (2008) Dietary intake of organophosphorus pesticide residues through vegetables from Kumasi, Ghana. Food Chem Toxicol 46(12):3703–3706 DOI: https://doi.org/10.1016/j.fct.2008.09.049
11. Davis MK, Boone J, Moran JE, Tyler JW, Chambers JE (2008) Assessing intermittent pesticide exposure from flea control collars containing the organophosphorus insecticide tetrachlorvinphos. J Exp Sci Environ EPI 18:564–570. DOI: https://doi.org/10.1038/sj.jes.7500647
12. EPA (2000) R.E.D. Facts: ethyl parathion. EPA. Washington DC. EPA-738-F00-009. https://www3.epa.gov/pesticides/chem_search/reg_actions/reregistration/fs_PC-057501_1-Sep-00.pdf
13. EPA (2017) Tetrachlorvinphos (TCVP). https://www.epa.gov/ingredients-used-pesticide-products/tetrachlorvinphos-tcvp
14. IARC (2017) "IARC Monographs on the evaluation of carcinogenic risks to humans. Some organophosphate insecticides and herbicides" volume 112
15. Lu J, Sun Y, Waterhouse GIN, Xu Z (2018) A voltametric sensor based on the use of reducted graphene oxide and hollow gold nanoparticles for the quantification of methy parathion and parathion in agricultural products. Adv Polym Technol 37:3629–3638
16. Richmond M (2018) Glyphosate: a review of its global use, environmental impact, and potential health effects on humans and other species. J Environ Stud Sci 8(4):416–434. https://doi.org/10.1007/s13412-018-0517-2
17. Schneider K (1991) "Deadly pesticide may face US Ban." NY Times. New York, NY
18. UTZ Certified (2015) List of banned pesticides and pesticides watchlist. Amsterdam, The Netherlands, UTZ Certified. https://utz.org/wp-content/uploads/2015/12/EN_UTZ_List-of-Banned-PesticidesWatchlist_v1.0_2015.pdf

Chapter 3
Environmental Presence of Parathion, Malathion, Diazinon, Tetrachlorvinphos, and Glyphosate

Abstract When pesticides and herbicides are applied to crops or livestock, employed in landscaping, or used domestically to control pests in residences or on domestic animals, they have the potential to spread in the environment. Some of this spread may be localized, found in micro-environments, but the extensive use of these chemicals can result in widespread environmental distribution: in soils, ambient air, and bodies of water. This distribution, in turn, has the potential for both acute and chronic health responses among humans, domestic animals, wildlife, and plants. The following chapter evaluates data from both laboratory and field studies that examine the capacity of each substance to bind onto soils, disperse in the air, enter streams and waterways, and, when applicable, enter the water table. It also augments these findings with additional environmental information, to assess the environmental spread of the four pesticides and glyphosate into the larger ecosystem. While none of the available data are a comprehensive global model, they do furnish tools for predictive models in areas where environmental information is less readily available.

3.1 Properties of Organic Phosphate Pesticides Influencing Environmental Distribution

In considering the environmental distribution of each of the organic phosphate pesticides that were considered by IARC in 2015, it is important not only to take into account the amount of use but also water solubility and soil binding properties. Each of the four organophosphates has limited water solubility (in mg/L ranges). However, soil binding coefficients[1] are significantly larger, although the degree of binding is influenced by the type of soil. Airborne distribution may be localized, especially when the pesticides are applied in mist form to a specific area. However, wider distribution can occur from the movement of pesticide-bound soil particles that become airborne. Finally, in looking at distribution, it is important to consider

[1] A numerical measure of the mobility of a substance in soil.

M. Richmond, *Cancer Hazards: Parathion, Malathion, Diazinon, Tetrachlorvinphos and Glyphosate*, AESS Interdisciplinary Environmental Studies and Sciences Series,
https://doi.org/10.1007/978-3-030-81953-8_3

the half-life of each pesticide, whether the breakdown products of each are toxic and the nature of their distribution.

Three of the four pesticides under consideration, parathion, malathion, and diazinon, have different chemical groups attached to the thiophosphate ester; however, they share the thiophosphate ester as a common structure. Tetrachlorvinphos is an oxygen ester. This is an important consideration in environmental distribution because parathion, malathion, and diazinon especially when airborne can undergo photo-oxidation, becoming the oxygen analog, a more potent choline esterase inhibitor. This does not apply to tetrachlorvinphos, already an oxygen ester. Photo-oxidation and hydrolysis are shown in Fig. 3.1.

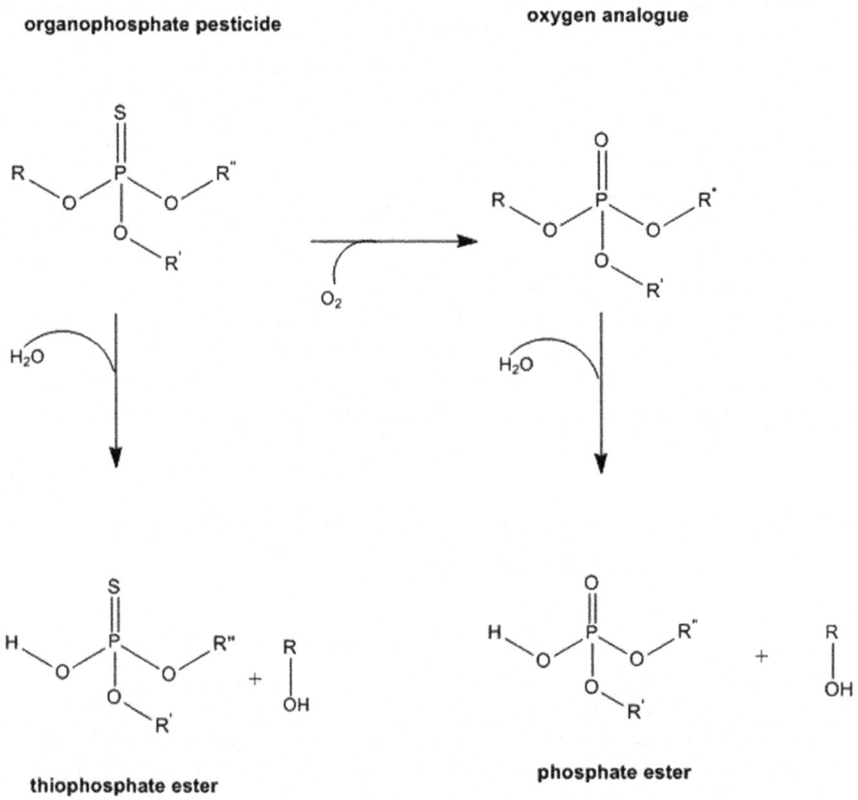

Fig. 3.1 Reactions involved in the oxidation of generic organophosphate pesticides and their subsequent hydrolysis

3.1.1 Parathion

As noted earlier, the use of parathion has been banned or severely restricted in a large number of countries and geographic unions. It is reasonable to expect that environmental distribution in these locales will be relatively small. At the same time, verifiable use of data is missing for several other countries. Hence, it is important to consider the ways that parathion could be found in the environment.

The half-life of parathion depends on the pH of the environment in which it is located. Parathion is more stable under acid conditions, with a half-life of 272 days at a pH of 4, decreasing to a half-life of 130 days at a pH of 9 [18, 19]. On the other hand, the photo-degradation half-life of parathion on water surfaces is quite variable. It has been reported to be more rapid in some cases, with values ranging from 1 to 10 days [9] but a slow process in other cases [26]. One of the photo-degradation products found in significant amounts is the oxygen analog, paraoxon, a compound with higher acute toxicity [26]. As shown in the figure above, both can be hydrolyzed.

While parathion is not appreciably water-soluble, when found on water surfaces, it can become volatile and subsequently absorbed onto soils or sediments. Rates of volatilization vary and differ according to the depth of the water body [5, 6]. Parathion found in water can also be absorbed into soil and sediments.

Little specific information is available about the type of soils that best absorb parathion, but a large body of data is available confirming that it is readily absorbed [5, 6, 9]. Parathion can enter the water on soil sediments, but, as noted earlier, it is not per se water-soluble. In an article published in 2005, Salfih-Hdadi and coworkers [22] studying the kinetics of soil binding in laboratory settings found strong absorption of parathion by soil, with both parathion and its metabolites being found at the topsoil layers. Sorption lasted for long periods, suggesting that soil served as a reservoir for parathion residues.

3.1.2 Malathion

Because it has been used in significantly larger amounts than parathion, while some of the same environmental distribution parameters may apply, more malathion will be found in the environment. Of particular concern is the question of malathion environmental distribution in areas where it has been used extensively, such as some farming applications, or where its use has been both heavy and focused, as, for example, for mosquito or cotton boll weevil control. In addition to information about air distribution and malathion in water, a significant body of knowledge exists for malathion in soils, including binding kinetics, binding affinities of different soils, and implications of these differences.

Malathion application is generally through aerial spraying or ground fogging on crops or other areas for purposes such as mosquito abatement [3]. Residues and droplets can remain in the atmosphere or can enter through volatilization from the surface of sprayed plants and trees.

Although volatilization generally results in elevations of malathion in the region where the application has taken place, atmospheric conditions including wind and humidity may allow the vapors to spread into more distant environments. In California, where malathion has been used extensively in the Central Valley, it has been found in the elevated environment of the Sierra Mountains and Sequoia National Park [3, 17, 25]. Other bodies of water where spread has been found include the east coast of the Chesapeake Bay and the east coast of Florida [3, 17].

Like parathion, malathion deposition on soil varies according to the type of soil. Soil particles of tightly bound malathion can transport the pesticide to long distances [3, 17, 25]. For example, when it is bound tightly to kaolin or soils with a high degree of kaolin, it does not easily biodegrade, and with air currents, it will be transported over long distances [3, 25].

The breakdown of soil-bound malathion is predominantly through degradation by aerobic organisms. A significant product is malaoxon, the oxygen analog of malathion, which is more toxic than malathion. Time for decomposition depends upon the soil pH, the nature of the soil (e.g., sand or loam content) as well as the presence of organic matter bound to the soil. The pesticide's half-life in soils of higher acidity tends to be greater than that for more alkaline soils. An additional consideration is the amount of organic matter found in the soil. Time for transformation to malaoxon is slower in soils with a greater amount of organic matter [17]. Available data indicate that malathion in the soil is unlikely to photo-degrade, or to be released into groundwater [3].

While it is generally felt that pesticide from the soil is unlikely to enter groundwater, data do indicate that it can enter rivers and streams. Some of this deposition is the result of spraying with an insufficient buffer zone between the sprayed area and the body of water. Deposition can also occur as a result of the atmospheric movement of malathion-containing soil particles, while still other deposition can occur as a result of atmospheric drift, as, for example, the findings of the pesticide and its degradation products on the beaches and water on the east coast of Florida after coastal marsh spraying for mosquito abatement [17].

Breakdown in water is pH-dependent and also reflects the microbial community in the water [3, 17, 25]. Malathion breaks down relatively quickly at high pH values, but the breakdown is considerably slower at low values, where the microbial environment has been shown to play a more important role [3]. Agency for Toxic Substances and Disease Registry (ATSDR) reports that photo-degradation does not appear to play a significant role in the breakdown.

3.1.3 Diazinon

Diazinon, like malathion, has been found in air and aerosols, in soils, and in bodies of water. In countries where its use has been restricted, such as the US or UK where residential use is not allowed, the amounts in the environment have decreased over time. However, globally, it is still used in rice cultivation.

Diazinon can enter the atmosphere through volatilization or as mists following application. Under appropriate circumstances, the volatile form can reach high elevations. As is true of malathion, California has reported diazinon in the Sierra Nevada Mountains following agricultural application in the central valley [4]. Following application, it can also be found in rain [13].

While diazinon does bind to soil, it is considered to be relatively mobile, moving from soil to the atmosphere or bodies of water [4]. Binding, in turn, is dependent on several factors: the amount of organic material in the soil, the soil texture (e.g., loam versus sandy soil), and, to some extent, the soil pH. The release of diazinon from soils with a high degree of organic matter is less than in soils with low content. The release from sandy soils is also higher than from soils with other textures [4]. Breakdown through hydrolysis is reported to be higher in more acid soils [13], and occurs primarily through microbial degradation; while the ultimate breakdown product may be carbon dioxide, a breakdown intermediate is diazoxon, which is further degraded to oxypyrimidine, a more mobile soil component [13]. Diazinon in the soil can also be broken down through photolysis [13].

Because of its mobility, the pesticide can be found in surface water and may enter streams through run-off. It has also been reported in groundwater [4, 13]. Diazinon's use in rice cultivation warrants particular attention since rice cultivation requires water. To ensure that the developing plants have adequate water, farmers will often flood fields, and because developing rice plants are susceptible to a number of pests, many farmers also use pesticides to minimize pest damage. While data on pesticide use vary from country to country, not surprisingly, findings from several countries report significant amounts of diazinon in the fields and water bodies where rice is growing [1, 2, 12, 24]. Meaningful amounts can also be found in surface water [2] and may also enter the groundwater.

3.1.4 Tetrachlorvinphos

As noted in Chap. 2, the global use of tetrachlorvinphos has, since its introduction, become quite limited. The EU has banned its use and EPA has banned it for crops. It is still used to treat farm livestock in the US, and a final EPA decision regarding its re-registration for flea and tick control on domestic animals is still pending [10]. As of 2017, it was registered for use in Canada, South Africa, and Australia [11, 16].

Given the small amount of use globally, little detailed information is available about the environmental presence of tetrachlorvinphos. In the absence of such information, the presence of the pesticide is based more on chemical properties which are likely to be similar to those of the other three organophosphate pesticides evaluated by IARC. In the 2017 Monograph, IARC reported that tetrachlorvinphos was detected in 4% of dust samples collected from residences in Oakland, CA, and 10% of farmworker homes in Salinas, CA [16].

Although it is relatively insoluble in water (approximately 12 mg/L), tetrachlorvinphos breaks down slowly in water in a pH-dependent manner. The rate of hydrolysis increases with increasing pH [18, 19]. Little confirming data are available, but concern has been expressed that tetrachlorvinphos can enter waterways from domestic animals using flea collars or from treated livestock.

3.1.5 *Glyphosate*

Of the chemical substances considered in the 2017 IARC Monograph, glyphosate is by far the most widely used. Data developed by the US Geological Survey for agricultural use in 2012 indicated that, as compared with malathion (the most widely used pesticide evaluated in the Monograph), glyphosate use was orders of magnitude higher, approximately 140,000 US tons of glyphosate vs. 457 US tons of malathion [7]. It is not, therefore, surprising that the environmental presence of glyphosate has been more exhaustively studied and that its occurrence has been more completely characterized.

Glyphosate can enter the soil through direct application of the herbicide or from the exudate of plant root tips. Microorganisms quickly break glyphosate down into glycolic acid and aminomethylphosphoric acid (AMPA). The breakdown pathway is shown in Fig. 3.2.

Glycolic acid (glyoxylate), in turn, breaks down into carbon dioxide and water, whereas AMPA breaks down into phosphate and ammonia. While carbon dioxide (and possibly ammonia) may be released as gases, the remaining breakdown products, because they are polar under most soil conditions, generally become bound to soil

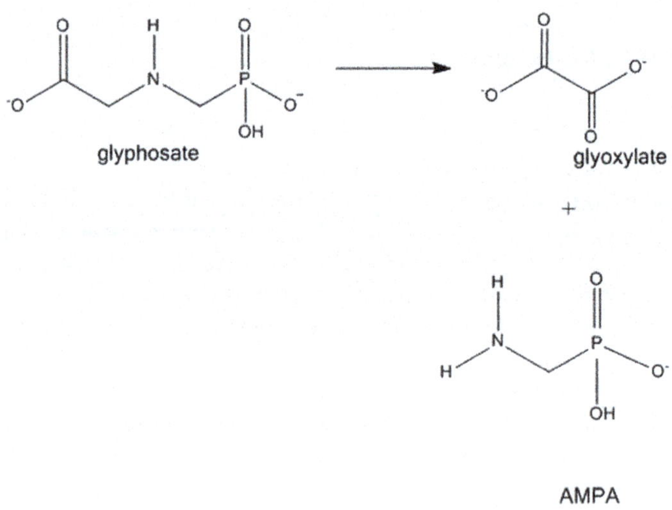

Fig. 3.2 Breakdown of glyphosate into glycolic acid (glyoxylate) and AMPA

[21]. As well as glyphosate binding to soil itself, an additional concern is binding of the phosphate which is released when AMPA breaks down [14].

Glyphosate and AMPA, both polar molecules, generally bind tightly to soils, with half-lives ranging from a few days to several months [15]. A number of factors determine the strength of binding. Included are mineral content, phosphate concentration, and pH. Because of their positive ionic charges, soil iron, calcium, and aluminum content facilitate glyphosate binding. In contrast, soil with high phosphate content lessens the strength of glyphosate binding. Glyphosate absorbs more easily to soils with low pH values, becoming less tightly bound to soils with high values. An additional consideration is the microbial and plant composition of the soil. Glyphosate can be broken down by soil microorganisms or assimilated by non-target plants [14]. Because glyphosate and phosphate are both negatively charged chemicals, they may compete for binding sites. If, through other sources (such as fertilizers), a soil has high phosphate content, glyphosate will bind less readily [14].

Because both glyphosate and its immediate breakdown product, AMPA, are water-soluble, it is not surprising that, under the appropriate circumstances, both could be found in such bodies of water as ditches, drains, rivers, streams, lakes, ponds, and wetlands. Initially, it was considered unlikely that glyphosate would also be found to any appreciable extent in groundwater. More recent findings, however, indicate that glyphosate has also been found in groundwater. In a USGS study done by Battaglin and coworkers [8], glyphosate and AMPA were found in groundwater as well as other water samples. The occurrence of the two substances suggested the relatively rapid conversion of glyphosate to AMPA. Specifically, only 2.3% of all water samples contained glyphosate alone, whereas AMPA was found without glyphosate in 17.9% of samples analyzed. The amounts of both in groundwater were not considered to pose threats to humans or wildlife. However, Rendón-von Osten and Dzul-Caamal, analyzing groundwater in maize(corn)-farming community in the western Yucatan peninsula found glyphosate to be present in groundwater samples, as well as urine samples taken from farmers in various municipalities in the farming community [20]. Glyphosate was reported as widely used in maize farming in the Yucatan peninsula.

Factors such as microbial activity and tightness of binding to soil can influence the amount of glyphosate that enters bodies of water, but an additional factor is precipitation. Sanchís and coworkers, analyzing 140 groundwater samples in Spain, reported that approximately 40% contained glyphosate with higher concentrations being found following periods of heavy precipitation that occurred after periods of drought. The authors suggested that this represented leaching from soil [23].

Glyphosate can enter the atmosphere when it is applied through spraying. However, uptake by plants and soils is rapid, and the herbicide does not enter the atmosphere from the soil. And, as noted earlier, it is quickly broken down into AMPA [6].

References

1. Abdullah AR, Bajet C, Matin MA, Nhan DD, Sulaiman AH (1997) Ecotoxicology of pesticides in the tropical paddy field ecosystem. Environ Tox Chem 16(1):59–70. https://doi.org/10.1002/etc.5620160106
2. Arjmandi R, Tavakol M, Shayeghi M (2010) Determination of organophosphorus insecticide residues in the rice paddies. Int J Environ Sci Tech 7(1):175–182
3. ATSDR, (2003) Toxicological profile for malathion. US Department of Health and Human Services, Public Health Service. Agency for Toxic Substances and Disease Registry. Atlanta, GA
4. ATSDR (2008) Toxicological profile for diazinon. US Department of Health and Human Services, Public Health Service. Agency for Toxic Substances and Disease Registry, Atlanta, GA
5. ATSDR (2019) Toxicological profile for parathion. US Department of Health and Human Services, Agency for Toxic Substances and Disease Registry, Atlanta GA
6. ATSDR (2020) Toxicological profile for glyphosate. US Department of Health and Human Services, Agency for Toxic Substances and Disease Registry, Atlanta GA
7. Baker NT, Stone W (2015) Estimated annual agricultural pesticide use for counties of the conterminous United States, 2008–12: , U.S. Geological survey data series. pp 907–916. https://doi.org/10.3133/ds907
8. Battaglin WA, Meyer M, Kuivila KM, Dietze JE (2014) Glyphosate and its degradation product AMPA occur frequently and widely in US soils, surface water, groundwater and precipitation. J Am Water Res Assoc 50(2):275–290. https://doi.org/10.1111/jawr.12159
9. Cooperative Extension Offices of Cornell University, M. S. U., Oregon State University, and University of California at Davis (1993) Parathion pesticide information profile. Extoxnet. Ithica, NY
10. EPA (2017) Tetrachlorvinphos (TCVP). https://www.epa.gov/ingredients-used-pesticide-products/tetrachlorvinphos-tcvp
11. Erikson B (2020) Will EPA ban hazardous pesticide in pet collars? Chemical and Engineering News, American Chemical Society. Accessed 24 April 2020
12. Ghassempour M, Mohammadkhah A, Najati F, Rajadzadeh M (2002) Monitoring of the pesticide diazinon in soil, stem and surface water of rice fields. Anal Sci 18:779–783. DOI: https://doi.org/10.2116/analsci.18.779
13. National Pesticide Information Center (2011) Diazinon Technical Fact Sheet. Oregon State University. npic.ace.orst.edu
14. Hebert M-P, Fugère V, Gonzalez A (2019) The overlooked impact of glyphosate use on phosphorus loading in agricultural watersheds. Front Ecol Environ 17(1):48–56. https://doi.org/10.1002/fee.1985
15. Henderson AM, Gervais JA, Luukinen B, Buhl K, Stone D, Cross A, Jenkins J (2010) "Glyphosate general fact sheet; National Pesticide Information Center," Oregon State University Extension Services. http://npic.orst.edu/factsheets/glyphogen.html
16. IARC (2017) "IARC Monographs on the Evaluation of Carcinogenic Risks to Humans. Some Organophosphate Insecticides and Herbicides," Volume 112
17. Newhart K (2006) Environmental fate of Malathion. California Environmental Protection Agency, Department of Pesticide Regulation, Environmental Monitoring Branch. https://citeseerx.ist.psu.edu/viewdoc/download?doi=10.1.1.597.1678&rep=rep1&type=pdf
18. PubChem [Internet]. Bethesda (MD): National Library of Medicine (US), National Center for Biotechnology Information; 2004. PubChem Compound Summary for CID 991, Parathion [cited 2021 Sept. 27]. Available from: https://pubchem.ncbi.nlm.nih.gov/compound/Parathion
19. PubChem [Internet]. Bethesda (MD): National Library of Medicine (US), National Center for Biotechnology Information; 2004. PubChem Compound Summary for CID 5284462, Tetrachlorvinphos [cited 2021 Sept. 27]. Available from: PubChem [Internet]. Bethesda (MD): National Library of Medicine (US), National Center for Biotechnology Information;

2004. PubChem Compound Summary for CID 5284462, Tetrachlorvinphos [cited 2021 Sept. 27]. Available from: https://pubchem.ncbi.nlm.nih.gov/compound/Tetrachlorvinphosttps://pubchem.ncbi.nlm.nih.gov/compound/Tetrachlorvinphos
20. Rendon-von Osten J, Dzul-Caamal R (2017) Glyphosate residues in groundwater, drinking water and urine of subsistence farmers from intensive agriculture localities: a survey in Hopelchén, Campeche, Mexico. Int J Environ Res Public Health 14(5). Rendon-von Osten J, Dzul-Caamal R (2017). DOI: https://doi.org/10.3390/ijerph14060595
21. Richmond M (2018) Glyphosate: a review of its global use, environmental impact, and potential health effects on humans and other species. J Environ Stud Sci 8(4):416–434. https://doi.org/10.1007/s13412-018-0517-2
22. Saffih-Hdadi K, Bruckler L, Francois L, Barriuso E (2005) A model for linking the effects of parathion in soil to its degradation and bioavailability kinetics. J Environ Qual 35(1):253–267. DOI: https://doi.org/10.2134/jeq2005.0059
23. Sanchís J, Kantiani L, Llorca M, Rubio F, Ginebreda A, Fraile J, Garrido T, Farré M (2012) Determination of glyphosate in groundwater samples using an ultrasensitive immunoassay and confirmation by on-line solid-phase extraction followed by liquid chromatography coupled to tandem mass spectrometry. Anal Bioanal Chem 402(7):2335–2345. DOI: https://doi.org/10.1007/s00216-011-5541-y
24. Stadlinger N, Berg H., VandenBrink PJ, Tam NT, Gunnarsson JS (2018) Comparison of predicted aquatic risks of pesticides used under different rice-farming strategies in the Mekong Delta, Vietnam. Environ Sci Pollut Res Int 25(14):13322–13334. DOI: https://doi.org/10.1007/s11356-016-7991-4
25. Tchounwou PB, Patlolla A, Yedjou CG, Moore PD (2015) Environmental exposure and health effects associated with malathion toxicity. Toxicity and hazard of agrochemicals, Intech open, pp 71–91. DOI: https://doi.org/10.5772/60911
26. Wu C, Linden K (2008) Degradation and byproduct formation of parathion in aqueous solutions by UV and UV/H2O2 treatment. Water Res 42(19):4780–4790. Published online 2008 Sep 3. DOI: https://doi.org/10.1016/j.watres.2008.08.023

Chapter 4
Toxicology of Parathion, Malathion, Diazinon, Tetrachlorvinphos, and Glyphosate

Abstract While the mode of pesticide or herbicide action provides information about the acute toxicity of each agent to target species, a full understanding of the potential effects of each requires a greater understanding of acute toxicity as well as human metabolism and pharmacokinetics of each. To provide a context for a more detailed consideration of chronic health effects, the concept of body burden, that is, the amount of a particular chemical present and retained in a human or animal body and its implications to health from long-term or protracted exposure is also critical. These concepts are developed in the following chapter.

4.1 General Considerations Regarding Acute Pesticide Toxicology of Organophosphates

Intake of parathion, malathion, diazinon, and tetrachlorvinphos can be through dermal contact, through inhalation, and, orally either accidentally or through intention. While data for metabolic pathways from the dermal or respiratory intake is limited, animal study data of oral intake indicate that the pesticides are transported to the liver through the hepatic portal system where they can undergo a number of transformations. The specific transformations depend, in part, on the structure of the group attached to the phosphosulfur portion of the molecule. However, as is true with other liver transformations, oxidation is a primary process. This is of especial importance since the conversion of the phosphosulfur compound to the more toxic oxy analog is essential for the full toxic effects.

All four pesticides are neurotoxins, and as discussed in Chap. 1, their primary toxic action is inhibition of cholinergic neurotransmission. Cholinergic neurotransmission can be classified as muscarinic, that is, stimulated by muscarine, or nicotinic, that is, stimulated by nicotine. The neurotransmission process is itself associated with the central nervous system as well as autonomic ganglia, smooth muscles, and other parasympathetic organs involved in homeostasis. Organophosphate poisoning includes a number of muscarinic symptoms including salivation, lacrimation, urination, diarrhea, gastrointestinal symptoms, and emesis (nausea and vomiting). These

M. Richmond, *Cancer Hazards: Parathion, Malathion, Diazinon, Tetrachlorvinphos and Glyphosate*, AESS Interdisciplinary Environmental Studies and Sciences Series,
https://doi.org/10.1007/978-3-030-81953-8_4

processes are often referred to as SLUDGE. Other muscarinic symptoms include excess production of watery sputum, bronchospasm, abnormally slow heart, and constricted pupils. Poisoning can also involve nicotinic symptoms such as involuntary muscle twitches and muscle weakness. Some symptoms appear quickly; others appear sometime after exposure.

The binding of an organophosphate pesticide inhibitor can be reversed by treatment with pharmacologic agents. The primary treatment for acute poisoning is considered to be the treatment with atropine, which totally inhibits acetylcholine binding at muscarinic synapses. As a secondary treatment, pralidoxime chloride, sometimes referred to as 2-PAM is also used. Treatment with 2-PAM removes the pesticide phosphoryl group from choline esterases, restoring activity. However, if the "aging" process has occurred, the phosphoryl group remains tightly bound, and 2-PAM treatment is not effective. Treatment with 2-PAM initially evolved from in vitro studies, and a number of authorities consider it to be controversial [4, 6].

While not as non-polar as chlorinated hydrocarbon pesticides all organophosphate pesticides are lipid-soluble. Data indicate that amounts can accumulate in fat and liver tissue, in the brain, and in kidneys [1]. Transplacental transfer is also possible [9]. While detailed information regarding specific organophosphates is not readily available, it has been possible to identify metabolites common to parathion, malathion, diazinon, and tetrachlorvinphos in spot urine tests, and use this information to calculate the overall body burden among various age groups. Payne-Sturgis and coworkers [8] used information from CDC's National Health and Nutrition Examination Survey (NHANES) for the period of 1999–2002 to estimate body burden for various age groups. Their findings indicate that children in the age group 6–11 carry higher body burdens than any other age group, and these have been found to be especially high for this group of children in Monterey County, California, a major farming district.

4.2 Toxicology and Toxicokinetics of Parathion, Malathion, Diazinon, and Tetrachlorvinphos

4.2.1 Parathion

As noted above, following intake, like all organophosphate pesticides significant amounts of parathion are transported to the liver where parathion undergoes a number of metabolic transformations. The P450 cytochrome isoenzymes are involved in these transformations. Transformations involve both activation and oxidative transformation to paraoxon (the more toxic metabolite), as well as detoxification by esterases: parathion to the diethylthiophosphoric acid and p-nitrophenol, or paraoxon to diethylphosphoric acid and p-nitrophenol. The hydrolysis products, undergoing conjugation, can be excreted in urine. Either parathion or the transformation product paraoxon can bind via the alkylphosphate group to acetylcholine esterases. As noted, this specific binding is reversible, but if paraoxon becomes "aged" through the removal of an alkyl group from the phosphate, the inhibitor is irreversibly bound.

P450 isoenzymes are a large group of liver enzymes whose specific role is to metabolically transform xenobiotics, that is "foreign compounds" into forms that are more easily conjugated (essentially joined to "carrier" molecules) to move to the kidney and become excreted. These isoenzymes are a large family of proteins. Members of each family can be induced or repressed, depending on a number of circumstances. Some factors are related to age, gender, or other specific biological characteristics such as pregnancy. Others are more temporarily related to such issues as dietary status, general liver health, or the presence of other xenobiotics, including pharmacological agents. Hence, while it is relatively easy to describe the overall process of transformation, the rate at which various transformation and conjugation reactions occur or the yield of different products can vary from individual to individual, and a range of toxic responses can be expected.

4.2.2 *Malathion*

As is true for parathion, malathion is primarily transformed through the P450 cytochrome isoenzymes and conjugating enzymes in the liver. An important transformation to the more toxic malaoxon is catalyzed by a P450 cytochrome isoenzyme as are several hydroxylation reactions.

The group attached to the alkyl thiophosphate or alkylphosphate of either malathion or malaoxon (R group) can undergo carboxyesterase hydrolysis. In the case of malathion, the products are malathion monocarboxylic acid and/or malathion dicarboxylic acid; for malaoxon, the products are malaoxon mono- and malaoxon dicarboxylic acid. As is true of parathion or paraoxon, the phosphate ester bond to the R group may also be broken, yielding thiophosphoric acid or phosphoric acid esters. The malathion or malaoxon products are O,O-dimethyl thiophosphoric acid and O,O-dimethylphosphoric acid, respectively. Malaoxon and malathion can also bind reversibly to acetylcholine esterases, or it can undergo dealkylation (aging) and bind irreversibly [1].

As is true of parathion, the extent to which all of these reactions occur and which may predominate is dependent on the activities of the various P450 and conjugation enzymes that are involved in the reactions. And, as is generally the case, activities will, in part, depend on such intrinsic factors as gender, developmental stage, and age, or on extrinsic factors such as general health, dietary status, competition of other xenobiotics, and the presence of pharmacological agents.

4.2.3 Diazinon

The basic structural motif of diazinon is similar to that of parathion and malathion, namely, an R group attached to an alkyl thiophosphate group. It is, therefore, not surprising that the transformation and conjugation reactions have a similar pattern.

Diazinon, like the compounds previously discussed, can undergo oxidation to diazoxon, catalyzed by cytochrome P450 isoenzymes in the liver. As pointed out by ATSDR [2], while diazoxon is the more toxic compound, diazinon is, nevertheless, a relatively toxic compound.

The intermediate in the oxidation process can undergo spontaneous hydrolysis to form either diethylthiophosphoric acid (DETP) or diethylphosphoric acid (DEP) and 2-isopropyl-4-methyl-6-hydroxypyrimidine (IMHP,) which is a specific marker for exposure and intake of diazinon. Alternatively, the intermediate in the oxidation of diazinon can be transformed to diazoxon. Diazoxon can bind reversibly to acetylcholine esterases, or irreversibly the following aging. Diazoxon can also undergo hydrolysis to form IMHP and DEP. The compounds IMHP, DEP, and DETP are water-soluble and can be excreted in urine.

While the liver is the primary organ of transformation for diazinon, other aspects of diazinon transformation appear to occur in other body systems, including the brain, diaphragm, and blood [2]. Included in this process is the aging of diazoxon.

4.2.4 Tetrachlorvinphos

Less data regarding human metabolic transformations and the elimination of tetrachlorvinphos transformation products are available. Much available information comes from studies on a variety of animal species, including several that have agricultural importance. Unique to the four organophosphate pesticides under consideration, phosphorus in tetrachlorvinphos is a phosphate, not thiophosphate tri-ester. Thus, no metabolic conversion from a thiophosphate ester to a phosphate ester is required for its toxicological activity.

Based on available information from animal studies, an IARC working group participating in developing information on tetrachlorvinphos for the 2017 IARC monograph established a metabolic model for transformation that involves direct hydrolysis of the phosphate ester bond between the diethylphosphate bond and the tetrachloro-containing R group. The ultimate products of this process are an esterase-bound or free diethylphosphate and 2,4,5-trichlorophenacyl chloride. One of the methyl groups attached to the dialkylphosphate portion of tetrachlorvinphos can undergo dealkylation in a cytochrome P450 catalyzed reaction. The subsequent dealkylation product, can, in turn, undergo hydrolysis to yield monoalkylphosphate and 2,4,5-trichlorophenacylchloride. Many of the metabolites are then conjugated and the conjugates are excreted in urine [5].

4.3 Glyphosate

Unlike the four pesticides under consideration, glyphosate is water-soluble. Thus, while it may not be readily taken in through dermal exposure, it can enter the body through inhalation or ingestion (accidental or intentional). At the same time, because of its solubility, it is generally thought that glyphosate and its most likely metabolite, AMPA, are readily excreted. It should be noted that glyphosate, as well as occurring as the free acid, is also present in products as salt, including isopropylamine, sodium, and ammonium salts. In addition, many products containing glyphosate also include "inert" materials such as surfactants to facilitate absorption by plants. Given that the chemical make-up of many of the "inert" materials is proprietary, it is difficult to ascertain whether these materials could increase glyphosate intake into higher animals, or, in other ways, influence its metabolic conversions and toxicity.

4.3.1 Toxicology and Toxicokinetics of Glyphosate

Glyphosate that is taken in through digestion passes through the GI tract and is excreted in an unchanged form in feces. When taken in through inhalation or dermal exposure, it has been reported in the kidney, liver, brain tissue, and blood. However, little specific information is available as to the specific site or sites of metabolic transformation and translocation. The metabolic transformation of glyphosate was shown earlier, in Chap. 3.

Both glyphosate and its metabolic transformation product, AMPA, are polar and, therefore, water-soluble. While both are excreted in urine, the amount of glyphosate excreted is significantly larger. One study examining intentional ingestion of large amounts of glyphosate reported ratios of glyphosate to AMPA of roughly 140–150/1 in serum and blood. While results individual findings were variable, a recent study examining the urinary distribution of glyphosate and AMPA in urine samples taken from workers in a Chinese facility producing glyphosate report similar ratios of glyphosate to AMPA [10]. A review article by Niemann and coworkers [7] looking at several studies measuring glyphosate excretion in urine generally found that urinary concentrations of AMPA were significantly less than glyphosate concentrations. In the few findings of higher AMPA concentrations, the authors suggest that the AMPA concentrations may reflect the plant metabolism of genetically modified crops.

It is generally held that urinary excretion is the best measure of the body burden of glyphosate. Studies such as those cited all find that peak urinary excretion of glyphosate when measured on sequential days following the herbicide application is found in samples collected one day following exposure with a decline in the subsequent two days following. It should be noted that the studies cited have generally used a proscribed protocol of urine collection the day prior to application, followed by 24-h urine collection on days following, and that glyphosate concentrations in the day preceding application were below limits of detection. However, such studies do not address the question of multiple-day applications at different locations which can increase the body burden.

References

1. ATSDR (2003) Toxicological profile for malathion. US Department of Health and Human Services, Public Health Service. Agency for Toxic Substances and Disease Registry. Atlanta, GA. p. 129
2. ATSDR (2008) Toxicological profile for diazinon. US Department of Health and Human Services, Public Health Service. Agency for Toxic Substances and Disease Registry, Atlanta, GA., p.100
3. ATSDR (2019) Toxicological profile for glyphosate draft for public comment Atlanta, GA, ATSDR/CDC
4. Blumenberg A, Benabbas R, deSouza IS, Conigliaro A, Paladino L, Warman E, Sinert R, Weiner SW (2018) Utility of 2-pyridine aldoxime methyl chloride (2-pam) for acute organophosphate poisoning: a systematic review and meta-analysis. J Med Toxicol 14(1):91–98. DOI: https://doi.org/10.1007/s13181-017-0636-2
5. IARC (2017) "IARc monographs on the evaluation of carcinogenic risks to humans. Some Organophosphate Insecticides and Herbicides" Volume 112
6. Naughton SX, Terry A (2018) Neurotoxicity in acute and repeated organophosphate exposure. Toxicology 408(Sept 1):101–112. DOI: https://doi.org/10.1016/j.tox.2018.08.011
7. Niemann L, Sieke C, Pfeil R, Solecki R (2015) A critical review of glyphosate findings in human urine samples and comparison with the exposure of operators and consumers. J Verbraucherschutz Leb 10:3–12. https://doi.org/10.1007/s00003-014-0927-3
8. Payne-Sturges D, Cohen J, Castorina R, Axelrad DA, Woodruff TJ (2009) Evaluating cumulative organophosphorus pesticide body burden of children: a national case study. Environ Sci Technol 43:7924–7930. DOI: https://doi.org/10.1021/es900713s
9. PubChem [Internet]. Bethesda (MD): National Library of Medicine (US), National Center for Biotechnology Information; 2004-. PubChem Compound Summary for CID 991, Parathion; [cited 2021 Sept. 29]. Available from: https://pubchem.ncbi.nlm.nih.gov/compound/Parathion
10. Zhang F, Xu Y, Liu X, Pan L, Ding E, Dou J, Zhu B (2020) Concentration Distribution and Analysis of Urinary Glyphosate and Its Metabolites in Occupationally Exposed Workers in Eastern China. Int J Environ Res Public Health 17(8):2943. DOI: https://doi.org/10.3390/ijerph17082943

Chapter 5
Toxic Effects of Pesticides or Herbicide on the Environment and Environmental Species: Wildlife, Including Insects, Aquatic Animals, and Plants

Abstract A sentinel measure of the environmental spread and the persistence of parathion, malathion, diazinon, tetrachlorvinphos, and glyphosate gives data about the toxic effects on animals and plants in the larger ecosystem. However, a more complete understanding of environmental spread and its effects comes not only from environmental information but also from findings of controlled exposure laboratory studies. The following chapter examines information from both types of investigation. The focus is on organisms, including microorganisms found in the environment, including plant species, but will also address the issue of organism resistance.

5.1 Organophosphate Pesticides

5.1.1 Overall Considerations

The body of information about the effects of pesticides is vast, and it would be challenging to list all insects and other organisms that pesticides may harm. However, certain organisms, many essential to the ecosystem, are often found to react adversely, raising concerns about the widespread use of organophosphates with respect to overall ecosystem health. In some cases, their effects are more targeted, such as the use of malathion to control mosquito vectors for such conditions as the Zika virus. The following narrative addresses these issues and provides information regarding the specific effects of each.

5.1.1.1 Parathion

Not surprisingly, given that its use has been banned for some time in a large number of countries, parathion's effect on environmental organisms is not especially widespread. Data from the Extension Toxicology Network Pesticide Information Profiles updated in 1996 [13] noted that parathion was highly toxic to a number of

M. Richmond, *Cancer Hazards: Parathion, Malathion, Diazinon, Tetrachlorvinphos and Glyphosate*, AESS Interdisciplinary Environmental Studies and Sciences Series,
https://doi.org/10.1007/978-3-030-81953-8_5

bird species, "moderately toxic to fish and aquatic invertebrates" as, for example, crayfish, worms or snails, and toxic to honeybees. Little other information is available.

5.1.1.2 Malathion

Malathion is toxic to mosquitos, including mosquito vectors for several tropical diseases such as Zika, dengue, and malaria. In theory, if used judiciously and sparingly malathion would not contaminate the larger environment, not pose a significant problem, and address serious public health issues. However, at the same time, findings from studies, both conducted in the laboratory and the environment, suggest that malathion does affect a number of species, both animal and plant, and that control of its environmental spread can be challenging.

Microbial species in the environment are not affected by malathion; in fact, it is broken down environmentally by microbes [3]. However, other species, some important to maintaining a healthy ecosystem, are affected. Malathion is reported to be highly toxic to bees and some species of fish [11]. Depending on the temperature of the water, it can affect salmon, increasing the rate of disease as the water temperature increases. As found in laboratory studies, negative effects on daphnia are also temperature-dependent, which increase as the water temperature increases [39]. In contrast, PubChem reports that malathion has relatively low toxicity for a number of bird species.

Given the importance that earthworms have for soil health, it is not surprising that many studies have been published regarding the effects of malathion on earthworm health. In general, these have been laboratory studies, allowing investigators to observe various endpoints of exposure. A report published in 1992 compared the effects of a "single agricultural dose" of malathion and four times the agricultural dose on the population, bio-productivity, and metabolism of an earthworm species in India. Results indicated, within the parameters used, that increased exposure to malathion decreased measures of each of these endpoints [42]. A more recent study, also from India, administering different dose levels and treating groups for different time periods measured body weight, number of cocoons, and hatching efficiency of the cocoons. Findings indicated a significant dose–response as well as an effect from the duration of exposure [44]. A study conducted in Chile also looked at these same endpoints as well as effects on the male reproductive organ and the presence of altered (fragmented) DNA. Earthworm weight decreased with an increasing concentration of malathion treatment in a dose-dependent manner; treatment also affected the male reproduction capacity and altered DNA [12].

However, as noted, malathion has been seen as a valuable tool to control the growth of insect vectors such as those responsible for Zika, dengue, and malaria. Unfortunately, however, it is also not surprising, given the rapid growth and reproduction rates for these mosquitos that malathion and other pesticides could exert selective pressure for resistant strains. A recent publication from Brazil, where dengue is a serious illness, confirms this expectation. In the study, the *A. aegypti* larvae population were hatched synchronously. The adult insects, all of the same generation, were then treated with sufficient malathion to kill approximately 80% of the population. The same procedure was used with subsequent generations until malathion treatment was seen to kill only 20%, and this population was then maintained. This level of resistance was seen to continue for five generations [46]. While this may not replicate field conditions, it is a dramatic demonstration of the development of pesticide resistance in a targeted insect vector. A 2008 report on resistance development in Mediterranean fruit flies in Spain also examined the molecular biology foundations for resistance [30].

The presence of malathion (and other pesticides), by altering the ecosystem, can remove food sources for other organisms. An example of this phenomenon was reported by Groner and Relyer [17] examining the effects of malathion on aquatic communities. In the study, concomitant with the decreases in zooplankton and phytoplankton from malathion treatment, the number of tadpoles also decreased.

Although less has been reported about the effect of malathion on plant populations, a recent report described its effects on the growth of onion roots. A laboratory study, which used different concentrations of malathion, found that negative effects, both on root growth and onion DNA, occurred in a dose-dependent manner [44].

5.1.1.3 Diazinon

In many respects, the effects of diazinon on organisms in the ecosystem are similar to those of malathion. Unlike malathion, which has limited and highly regulated use for control of some mosquito populations, diazinon was, at one time, marketed as a chemical that could be used to control a large number of insect species, including invasive ants, termites, and carpenter ants. However, its residential use has now been canceled by EPA, and only commercial pest control agents can apply diazinon for insect control.

That diazinon is rapidly broken down in soil has led to a number of studies [1, 9], both to identify specific bacterial strains and to consider the possibility of using bacteria for remediation of soils containing the pesticide. At the same time, the half-life of diazinon in various non-sterile soil samples has been reported to vary over a period of 1–5 weeks [19] making such treatments challenging.

Diazinon has been reported to be toxic to a number of aquatic species, including such economically important fish as salmon and bluegill sunfish [25]. However, toxicity to fish, whether fresh or salt-water species, ranges in terms of toxicity depending on the species [19]. Diazinon also has a range of toxicity for other species but is reported to be less toxic to amphibians [19]. Freshwater snails also appear to be relatively resistant to diazinon's effects. A recently published and compelling article by Becker and coworkers [5] reports findings from field and laboratory studies in various areas of Lake Victoria in Kenya demonstrating that diazinon pollution has resulted in an increase in freshwater snails associated with the development of schistosomiasis.

Diazinon is found to be highly toxic to beneficial insects such as bees and invertebrates such as earthworms [19]. It is also reported to be very harmful to many bird species, including mallard ducks and has been linked to "bird kills following outdoor applications" [19].

Resistance to diazinon by formerly sensitive insect populations has also been noted. This includes the development of resistance in the horn fly, a cattle parasite [22, 27, 28] and resistance in two species of cattle tick [27].

5.1.1.4 Tetrachlorvinphos

Less frequently used than either malathion or diazinon, tetrachlorvinphos, nevertheless, does pose concerns regarding ecotoxicity. While only moderately toxic to a number of bird species, including mallard ducks, ring-necked pheasants, Japanese Quail, blue jays, house sparrows, and cardinals, tetrachlorvinphos is highly toxic to certain insects, including the leaf-cutter bee, honey bee, and the alkali bee [34]. Toxicity to fish varies. Tetrachlorvinphos is moderately harmful to spot but highly toxic to rainbow trout, blue gill, sunfish, and catfish; it is also highly toxic to Daphnia [34].

Tetrachlorvinphos resistance has been reported among house flies in New York state dairies [23] and in the lesser mealworm, a pest in poultry raising facilities [18].

5.2 Glyphosate

5.2.1 Overall Considerations

Given its widespread use, it is not unexpected that a large body of literature has been published about glyphosate's effects on the ecosystem. This ranges from research findings regarding effects on the rhizosphere, effects on aquatic organisms, insects, and the development of glyphosate resistance in weeds. Each of these will be addressed separately.

5.2.1.1 Glyphosate's Effects on the Rhizosphere

The rhizosphere can be defined as the area around a plant root that contains chemicals released by the root. These chemicals will, in turn, influence the population of microorganisms (bacteria and fungi) that exist in the immediate area. Because glyphosate inhibits the shikimate pathway involved in the synthesis of three amino acids essential for protein formation, in theory, plants, bacteria, and other microorganisms that require this pathway for growth would be unable to survive. The "bioavailability" of glyphosate does depend on the type of soil in which a glyphosate treated or resistant plant is growing; some soils will absorb more of the herbicide than the others. Nevertheless, when glyphosate is present and bioavailable in the rhizosphere, the inhibition creates selective pressure for bacteria and other microorganisms resistant to the herbicide's inhibitory effects, which may significantly alter the "normal" microorganism composition of the rhizosphere. At the same time, because glyphosate is a chelating agent, able to bind to a number of ions, it can also remove certain trace metal ions that are essential to good plant nutrition.

Understandably, most investigations have looked at glyphosate's effects in conditions that are carefully controlled, although they have been designed to simulate natural growth conditions. Zobiole and coworkers [47] detail the findings of a study examining uptake parameters from glyphosate-resistant soybeans treated with increasing rates of glyphosate application. The investigators found that as glyphosate rates of application increased, photosynthesis and macro- and micronutrient accumulation in leaf tissue decreased. They also found that nutrient uptake decreased. Newman and coworkers [37], studying rhizosphere effects on corn and soybean crops grown in a greenhouse over four growth periods, reported changes in the bacterial community found in rhizosphere soil. They noted that while rhizosphere soil of glyphosate-treated and non-treated contained a mix of Acidobacteria, Actinobacteria, and Proteobacteria, the proportion of Proteobacteria increased while the proportion of Acidobacteria decreased in the rhizosphere soil of glyphosate treated plants. The investigators pointed out the important role of Acidobacteria in the soil (in carbon and nitrogen utilization), and that the roles that this group of bacteria have may be altered because of glyphosate's ability to remove certain micronutrient ions.

Another study by the same group [36] involved genetic sequencing of bacterial species found in the rhizosphere soil of glyphosate-resistant corn and soybean grown over four growing seasons. They found that the bacteria located in the rhizosphere of treated soil had downregulated genes for carbohydrate, aspects of protein and amino acid metabolism, upregulated genes for respiration and other aspects of protein metabolism, and changes in genes regulating the uptake of several micronutrients. Glyphosate treatment of glyphosate-resistant corn and soybean was also reported to stimulate the colonization of roots by the fungal species *Fusarium* [24].

Glyphosate can be transferred from the roots of target to non-target plants in close proximity. While this may, in part, be a function of soil type, glyphosate, once in the rhizosphere, is sufficiently stable for transfer. Using the example of transfer from weed roots to tree roots in orchards, Neuman and coauthors [35] suggest glyphosate use needs to be reassessed.

5.2.1.2 Toxicity to Aquatic Animals

An earlier report [40] noted that a number of studies have examined the toxic effects of glyphosate on aquatic animals, including amphibians and fish. In general, these studies looked at toxic effects at or close to the LC_{50} of the organism under investigation, which was regarded as a concentration unlikely to be encountered in the environment. More recent studies have, however, looked at specific effects at concentrations of glyphosate or glyphosate in adjuvants more likely to be found in the environment or have examined in more detail parameters of glyphosate toxicity at concentrations close to the organism LC_{50}.

A recent study by Rodriquez and coworkers [41] examined the toxicity of glyphosate, its primary metabolite AMPA, an adjuvant, Atanor 48™, and a surfactant which is part of Atanor 48™. The specific focus was on the genotoxicity of these components for zebrafish larvae or rainbow trout gonads. Using the comet assay, a specific assay for DNA damage, findings indicated that all four compounds were genotoxic to zebrafish larvae, whereas only the surfactant present in the Atanor 48™ was genotoxic to rainbow trout gonads.

A recent study reported by Faria and coworkers [14] examines the effects on several neurological parameters of zebrafish following exposure to aquatic levels that are considered within the range of acceptable levels. According to the investigators, the "exposed fish presented a significant impairment of exploratory and social behaviors consistent with increased anxiety." While the results are clearly most relevant to findings regarding zebrafish, the authors speculate that "zebrafish is also a vertebrate model widely used in human neurobehavioral studies, [and] results are relevant not only for environmental risk assessment, but also for understanding the risk of chronic low-dose exposures on human health."

Various amphibians have also been studied, both in the laboratory and in the larger environment. Findings from the larger environment suggest that specific outcomes may be complex and subject to such other variations as food sources and environmental protections. A carefully designed field investigation conducted by Edge and coworkers [10] found that the outcomes for amphibian species to glyphosate's impact on wetlands are complex. While glyphosate has a negative impact on macrophyte distribution, as the distribution diminishes, other organisms that provide food and nutritional sources for larval amphibians increase. However, the population of larger predatory animals may also increase. Ultimately, the investigators concluded that glyphosate's overall impact is a function of the length of time it contaminates the wetland.

Recent laboratory findings also have focused on glyphosate's development effects on given amphibians. An investigation on its effects on the African clawed frog *Xenopus laevis* reported that exposure to anticipated ambient concentrations had a negative impact on oocyte development, which, in turn, would affect egg-laying and maturation [43]. A study reported by a Brazilian research group examined glyphosate's effects on two species of amphibians common to several Latin American countries, including Brazil, Argentina, and Paraguay [21]. Amphibians were exposed, in the laboratory, to aqueous concentrations of glyphosate permissible under the current Brazilian regulations. Investigators found that not only did glyphosate lead to death in animals studied, but that those surviving had a number of chronic issues, including a loss in expected mass and shorter growth length.

5.2.1.3 Effects on Insect Species

Glyphosate has been reported to affect a number of insect species and arthropods, including mosquitos, aphids, bees, and varieties of beetles [40]. A recent study looking at glyphosate's impact on infection of mosquitos by the malaria parasite reported that mosquitos were less likely to be infected if they had been exposed to "field realistic" doses of glyphosate. However, under nutritional stress, the glyphosate effect was not observed [4]. The authors attributed this finding to differences in digestive functioning under the two circumstances. A number of negative consequences of glyphosate exposure have also been reported for arthropod predators that are important for the control of other agricultural pests. This includes two types of spiders [40].

Several studies have examined the consequences of honey bee exposure to glyphosate. The flight pattern and homing time of honey bees are affected, while brood development was not reported to be altered [40]. Perhaps reinforcing earlier observations about mosquitos, the authors of a recent paper [33] found that glyphosate, as an inhibitor of the shikimate pathway found in bacteria, inhibited bacteria in the bee gut, changing the bacterial composition, which in turn can affect honey bee health. In a subsequent study, Motta and Moran noted that changes in the gut microbial make-up occur in a dose-dependent way. This includes duration and intensity, as well as timing of exposure [32].

A 2019 publication by Farina and coworkers [15] notes that glyphosate is found as a contaminant in honey, most likely because of drift which results in bees collecting nectar from glyphosate contaminated plants. Implicit in these findings is that beehives are also contaminated. To examine the effects of this contamination, the investigators measured a number of endpoints involved in colony health and reproductivity. Findings included an inhibited odor detection (whether nectar content was high or low in plants) after herbicide exposure; a dose-responsive decrease in homeward flight time; an inhibited olfactory "learning" process in young bees; and poorer brood development. Similar findings were reported by Luo and coworkers [29].

5.2.1.4 Effects on Earthworms

Findings from studies looking at glyphosate's effect on earthworms, while generally reporting negative effects, are complex. In part, this is because studies have used different earthworm species and have considered a variety of different endpoints.

A study of data collected over a period of more than 60 years reported that conventional tillage of soil decreases the abundance, biomass, and community structure of earthworms [6]. This finding has been used to argue that because glyphosate kills unwanted weeds it decreases the need for soil tillage which, in turn, would benefit the earthworm population. However, a number of investigations suggest that earthworms exposed to glyphosate can be adversely affected. In 2015 Gaupp-Berghausen and coworkers [16] reported that glyphosate based herbicides reduced reproduction and activity of earthworms, leading to localized concentrations of soil nutrients. A laboratory-based study involving earthworm exposure to soils containing varying amounts of glyphosate found that earthworms maintained in all soils for 56 days had 50% weight decreases [8].

Other studies have looked at a number of glyphosate's indirect effects on earthworms. A laboratory-based study reported by Van Hoesel and co-investigators [45] examined the effects on earthworms from "seed dressing," with glyphosate. Seed dressing is the practice of treating crop seeds before planting with given herbicides or pesticides. The investigators found that glyphosate exposure from the seed dressings did not have any appreciable effects on earthworms used in the study. An investigation reported by Lescano and coworkers [26] found that a particular species of earthworm, *Eisenia fedita,* was efficient in removing glyphosate and its immediate metabolite, AMPA from treated soils. In contrast, a Nigerian study looking at the indirect effects of glyphosate on earthworms, namely earthworm casts, reported that casts from three varieties of earthworm exposed to glyphosate reduced the growth and fruit production of tomatoes [38].

5.2.1.5 Glyphosate Resistance in Weeds

The growth of weeds resistant to glyphosate was first reported in 1996, and as of 2018, according to Heap and Duke, a total of 38 different weed species are resistant to glyphosate. These species are found in 37 countries made up of 34 crop-growing and six non-crop-growing environments [20]. Clearly, given the growth and the continuing use of glyphosate, it is likely that the number of resistant species has grown more since 2018. One approach to weed resistance to glyphosate is the additional use of other herbicides. A recent publication considering control of glyphosate resistance in a Mexican citrus orchard compared a number of herbicide treatments, examining glufosinate, alone or together with glyphosate. Glufosinate, a chemically similar compound, but with a different biochemical effect on plant growth was found to be, from an economic approach, the best alternative approach [2]. Other investigators have examined alternative cropping practices [7]. And, at the same time, industrial organizations have developed crop seeds with multiple herbicide resistance, encouraging the use of multiple herbicides to control weeds. In contrast, a publication by Mortensen et al. notes that "If herbicide-resistant-weed problems are addressed only with herbicides, evolution will most likely win," and suggests other strategies for weed management [31].

References

1. Abo-Amer AE (2011) Biodegradation of diazinon by Serratia marcescens DI101 and its use in bioremediation of contaminated environment. J Microbiol Biotechnol 21(1):71–80. DOI: https://doi.org/10.4014/jmb.1007.07024

2. Alcántara-de la Cruz R, Domínguez-Martínez P, da Silveira HM, Cruz-Hipólito HE, Palma-Bautista C, Vázquez-García JG, Domínguez-Valenzuela JA, De Prado R (2019) Management of glyphosate-resistant weeds in Mexican citrus groves: chemical alternatives and economic viability. Plants (Basel) Sept 4 8(9) 325. doi: https://doi.org/10.3390/plants8090325
3. ATSDR (2003) Toxicological profile for malathion. US Department of Health and Human Services, Public Health Service. Agency for Toxic Substances and Disease Registry. Atlanta, GA
4. Battaillard D, Christe P, Pigeault R (2020) Impact of field-realistic doses of glyphosate and nutritional stress on mosquito life history traits and susceptibility to malaria parasite infection. Ecol Evol 10:5079–5088. https://doi.org/10.1002/ece3.6261
5. Becker JM, Ganatra A, Kandie F et al (2020) Pesticide pollution in freshwater paves the way for schistosomiasis transmission. Sci Rep 10:3650
6. Broines MJI, Schmidt O (2017) Conventional tillage decreases the abundance and biomass of earthworms and alters their community structure in a global meta-analysis. Glob Chang Biol 23(10):4396–4419. https://doi.org/10.1111/gcb.13744
7. Colbach N, Fernier A, Le Corre V, Messean A, Darmency H (2017) Simulating changes in cropping practises in conventional and glyphosate-tolerant maize. I. Effects on weeds. Environ Sci Pollut Res 24:11582–11600. DOI: https://doi.org/10.1007/s11356-017-8591-7
8. Correia FV, Moreira J (2010) Effects of glyphosate and 2,4-D on earthworms (Eisenia foetida) in laboratory tests. Bull Environ Contam Toxicol 85(3):264–268. DOI: https://doi.org/10.1007/s00128-010-0089-7
9. Cycona M, Wójcik M, Piotrowska-Segetc Z (2009) Biodegradation of the organophosphorus insecticide diazinon by Serratia sp. and Pseudomonas sp. and their use in bioremediation of contaminated soil. Chemosphere 76(4):494–502. DOI: https://doi.org/10.1016/j.chemosphere.2009.03.023
10. Edge CB, Baker L, Lanctot CM, Melvin SD, Gahl MK, Kurban M, Navarro-Martin L, Kidd KA, Trudeau VI, Thompson DG, Mudge JF, Houlahan JF (2020). Compensatory indirect effects of an herbicide on wetland communities. Sci Total Environ 20:718
11. Editors of Encyclopedia Britanica (2014) Malathion insecticide. Britanica. E. Britanica
12. Espinoza-Navarro O, Bustos-Obregón E (2005) Effect of malathion on the male reproductive organs of earthworms, Eisenia foetida. Asian J Androl 7(1):97–101.DOI: https://doi.org/10.1111/j.1745-7262.2005.00005.x
13. EXTOXNET: Pesticide Information Project of Cooperative Extension Offices of Cornell University, O. S. U., the University of Idaho, and the University of California at Davis and the Institute for Environmental Toxicology, Michigan State University (1996) Pesticide Information Profiles: Parathion. E. T. Network, Oregon State University
14. Faria M, Bedrossiantz J, Ramirez JRR, Mayol M et al (2020) Glyphosate targets fish monoaminergic systems leading to oxidative stress and anxiety. Environ Int 106. DOI: https://doi.org/10.1016/j.envint.2020.106253
15. Farina WM, Balbuena M, Herbert LT, Mengoni Goñalons C, Vázquez DE (2019) Effects of the herbicide glyphosate on honey bee sensory and cognitive abilities: individual impairments with implications for the hive. Insects 10(10):354. DOI: https://doi.org/10.3390/insects10100354
16. Gaupp-Berghausen M, Hofer M, Rewald B, Zaller JG (2015) Glyphosate-based herbicides reduce the activity and reproduction of earthworms and lead to increased soil nutrient concentrations. Sci Rep 5(Aug 5)
17. Groner ML, Relyea R (2011) A tale of two pesticides: how common insecticides affect aquatic communities. Freshw Biol 56:2391–2404. https://doi.org/10.1111/j.1365-2427.2011.02667.x
18. Hamm RI, Kaufman P, Reasor CA, Rutz DA, Scott JG (2006) Resistance to cyfluthrin and tetrachlorvinphos in the lesser mealworm, Alphitobius diaperinus, collected from the eastern United States. Pest Manag Sci 62(7):673–677. DOI: https://doi.org/10.1002/ps.1230
19. National Pesticide Information Center (2011). Diazinon technical fact sheet. Oregon State University. npic.ace.orst.edu
20. Heap I, Duke S (2018) Overview of glyphosate-resistant weeds worldwide. Pest Manag Sci 74(5):1040–1049. DOI: https://doi.org/10.1002/ps.4760

21. Herek JS, Vargas L, Trindale SAR, Rutkoski CF, Macagnan N, Hartman PA, Hartmann MT (2020) Can environmental concentrations of glyphosate affect survival and cause malformation in amphibians? Effects from a glyphosate-based herbicide on Physalaemus cuvieri and P. gracilis (Anura: Leptodactylidae). Environ Sci Pollut Res Int 27(18):22619–22630
22. Holderman CJ, Swale D, Bloomquist JR, Kaufman PE (2018) Resistance to Permethrin, β-cyfluthrin, and Diazinon in Florida Horn Fly Populations. Insects 9(2). DOI: https://doi.org/10.3390/insects9020063
23. Kaufman PE, Scott J, Rutz DA (2001) Monitoring insecticide resistance in house flies (Diptera: Muscidae) from New York dairies. Pest Manag Sci 57(6):514–521. DOI: https://doi.org/10.1002/ps.319
24. Kremer RJ, Means N (2009) Glyphosate and glyphosate-resistant crop interactions with rhizosphere microorganisms. Eur J Agron 31(3):153–161
25. Larkin DJ, Tjeerdema R (2000) Fate and effects of diazinon. Rev Environ Contam Toxicol 166:49–82
26. Lescano MR, Masin C, Rodríguez AR, Godoy JL, Zalazar CS (2020) Earthworms to improve glyphosate degradation in biobeds. Environ Sci Pollut Res Int 27(21):270023–227031. DOI: https://doi.org/10.1007/s11356-020-09002-w
27. Li AY, Davey R, Miller RJ, George JL (2003) Resistance to coumaphos and diazinon in Boophilus microplus (Acari: Ixodidae) and evidence for the involvement of an oxidative detoxification mechanism. J Med Entomol 40(4):482–490. DOI: https://doi.org/10.1603/0022-2585-40.4.482
28. Li AY, Guerrero F, Almazan Garcia C, George JE (2003) Survey of resistance to permethrin and diazinon and the use of a multiplex polymerase chain reaction assay to detect resistance alleles in the horn fly, Haematobia irritans irritans (L.). J Med Entomol 40(6):942–949. DOI: https://doi.org/10.1603/0022-2585-40.6.942
29. Luo QH, Gao J, Guo Y, Liu C, Ma YZ, Zhou ZY, Dai PL, Hou CS, Wu YY, Diao QY, et al (2021) Effects of a commercially formulated glyphosate solutions at recommended concentrations on honeybee (Apis mellifera L.) behaviours. Sci Rep 11(1)
30. Maganana C, Hernández-Crespo P, Brun-Baraleb A, Couso-Ferrera F, Bride J-M, Castanera P, Feyereisen R, Ortegoa F (2008) Mechanisms of resistance to malathion in the medfly Ceratitis capitata. Insect Biochem Mol Biol 38(8):756–762. DOI: https://doi.org/10.1016/j.ibmb.2008.05.001
31. Mortensen DA, Egan J, Maxwell BD, Ryan MR, Smith RG (2012) Navigating a critical juncture for sustainable weed management. Bioscience 62(1):75–84. https://doi.org/10.1525/bio.2012.62.1.12
32. Motta EVS, Moran IN (2020) Impact of glyphosate on the honey bee gut microbiota: effects of intensity, duration, and timing of exposure. mSystems 5(4). https://doi.org/10.1128/mSystems.00268-20
33. Motta EVS, Raymann K, Moran NA (2018) Glyphosate perturbs the gut microbiota of honey bees. Proc Natl Acad Sci USA 115(41):10305–10310. DOI: https://doi.org/10.1073/pnas.1803880115
34. PubChem [Internet]. Bethesda (MD): National Library of Medicine (US), National Center for Biotechnology Information; 2004-. PubChem Compound Summary for CID 5284462, Tetrachlorvinphos; [cited 2021 Sept. 29]. Available from: https://pubchem.ncbi.nlm.nih.gov/compound/Tetrachlorvinphos
35. Neumann G, Kohls S, Landsberg E, Souza KS, Yamada T, Romheld V (2006) Relevance of glyphosate transfer to non-target plants via the rhizosphere. Zeitschrift für Pflanzenkrankheiten und Pflanzenschutz Sonderheft XX:963–969
36. Newman MM, Hoilett N, Lorenz N, Dick RP, Liles MR, Ramsier C, Kloepper JW (2016) Glyphosate effects on soil rhizosphere-associated bacterial communities. Sci Total Environ 543:155–160. DOI: https://doi.org/10.1016/j.scitotenv.2015.11.008
37. Newman MM, Lorenz N, Hlilet N, Lee NR, Dick RP, Liles MR, Ramsier C, Kloepper JW (2016) Changes in rhizosphere bacterial gene expression following glyphosate treatment. Sci Total Environ 553:32–41. DOI: https://doi.org/10.1016/j.scitotenv.2016.02.078

38. Owagboriaye F, Dedeke G., Bamidele J, Bankole A, Aladesida A, Feyisola R, Adeleke M, Adekunle O (2020) Wormcasts produced by three earthworm species (Alma millsoni, Eudrilus eugeniae and Libyodrilus violaceus) exposed to a glyphosate-based herbicide reduce growth, fruit yield and quality of tomato (Lycopersicon esculentum). Chemosphere 250:126270. https://doi.org/10.1016/j.chemosphere.2020.126270 (Epub 2020 Feb 20. PMID: 32105851)
39. PubChem [Internet]. Bethesda (MD): National Library of Medicine (US), National Center for Biotechnology Information; 2004. PubChem Compound Summary for CID 4004, Malathion; [cited 2021 Sept. 29]. Available from: https://pubchem.ncbi.nlm.nih.gov/compound/Malathion
40. Richmond ME (2018) Glyphosate: a review of its global use, environmental impact, and potential health effects on humans and other species. J Environ Stud Sci 8(4):416–434. https://doi.org/10.1007/s13412-018-0517-2
41. Rodrigues LB, Costa G, Tha EL, daSilva LR, et al (2019) Impact of the glyphosate-based commercial herbicide, its components and its metabolite AMPA on non-target aquatic organisms. Mutat Res/Genet Toxicol Environ Mutagen 842(June): 904–101. DOI: https://doi.org/10.1016/j.mrgentox.2019.05.002
42. Senapati BK, Biswal J, Pani C, Sahu SK (1992) Ecotoxicological effects of malathion on earthworms. Soil Biol Biochem 24(12):1719–1722
43. Slaby S, Titran P, Marchand G, Hanotel J, et al (2020) Effects of glyphosate and a commercial formulation Roundup®exposures on maturation of Xenopus laevis oocytes. Environ Sci Pollut Res 27:3697–3705. DOI: https://doi.org/10.1007/s11356-019-04596-2
44. Srivastava AK, Singh D (2020) Assessment of malathion toxicity on cytophysiological activity, DNA damage and antioxidant enzymesin root of Allium cepa model. Sci Rep 10(886). DOI: https://doi.org/10.1038/s41598-020-57840-y
45. Van Hoesel W, Tiefenbacher A, König N, Dorn VM, Hagenguth JF, Prah U, Widhalm T, Wiklicky V, Koller R, Bonkowski M, Lagerlöf J, Ratzenböck A, Zaller JG (2017) Single and combined effects of pesticide seed dressings and herbicides on earthworms, soil microorganisms, and litter decomposition. Front Plant Sci 21(8). DOI: https://doi.org/10.3389/fpls.2017.00215
46. Viana-Medieros PF, Bellinato DF, Valle D (2018) Laboratory selection of Aedes aegypti field populations with the organophosphate malathion: negative impacts on resistance to deltamethrin and to the organophosphate temephos. PLoS NeglTrop Dis 12(8):e0006734
47. Zobiole LHS, Junior RSDO, Kremer RJ, Muniz AS, de Oliveira Junior A (2010) Nutrient accumulation and photosynthesis in glyphosate resistant soybeans is reduced under glyphosate use. J Plant Nutr 33(12)

Chapter 6
Chronic Health Effects of Parathion, Malathion, Diazinon, Tetrachlorvinphos, and Glyphosate

Abstract While not all cancers result from environmental exposures, a significant number do. The development of cancer is not, however, the only response to chronic exposure. To understand the effects of chronic environmental exposure and, possibly, to identify vulnerable populations, it is helpful to look at other chronic exposure endpoints. The following chapter examines these endpoints, primarily in human populations, but also, as applicable, in higher animals that are environmentally exposed. Although cancer is the chronic health effect of concern in IARC's work, reports of other chronic conditions related to prolonged exposure to pesticides or herbicides, linkages between these conditions, and the possibility of the subsequent development of cancer are important considerations.

6.1 Effects of Organophosphates

As a group of chemical compounds, organophosphates have high acute toxicity and, in general, are not pervasive in the environment. However, relatively constant exposure to small "non-toxic" amounts of these same compounds can lead to some chronic health responses. Given their acute toxicity, the number of reports is relatively small and exposures are rarely to one specific organophosphate pesticide, making it difficult to specifically attribute a chronic effect to a given compound. Nevertheless, despite these limitations, several studies looking at developmental markers in children exposed in utero or during early childhood have found exposure-related deficits. While some of the studies report behavioral assessments of children shortly after birth, others have considered later times in childhood development.

Young and coworkers [16] report findings from a study that examined a number of neurobehavioral responses among neonates equal to or less than 2 months of age who were born to mothers working in the Salinas Valley agricultural community. The infants were exposed to organophosphate pesticides both in utero and shortly after birth. Prenatal exposure was determined by the mother's urinary levels of organophosphate pesticides and metabolites. While the specific organophosphate pesticides were not identified, given the pattern of use in the Salinas Valley, it is highly

M. Richmond, *Cancer Hazards: Parathion, Malathion, Diazinon, Tetrachlorvinphos and Glyphosate*, AESS Interdisciplinary Environmental Studies and Sciences Series,
https://doi.org/10.1007/978-3-030-81953-8_6

likely that exposure involved both malathion and diazinon. The neonates were found to score lower on a scale of neurobehavioral measures, and the decreases in scores were generally correlated with the level of urinary metabolites of the mother.

A study reported by Eskenazi and co-investigators [2] looked at neurobehavioral outcomes in children up to 24 months in age who were born to mothers exposed in utero to organophosphates while working in the Salinas Valley agricultural community. Urinary samples were measured for non-specific metabolites of organophosphate pesticides as well as metabolites specific to malathion and diazinon, both in the mothers and in the children. Prenatal exposure was determined by urinary levels found in samples provided by the mother. While the results were not consistent across all measures of neurobehavioral development, in general, as exposure to organophosphates increased (as determined by metabolites not specific to a given organophosphate pesticide), the positive measures of neurobehavioral development decreased.

Investigators in Ecuador examined neurobehavioral outcomes among children born to women exposed to organophosphate pesticides in the flower industry. Adjusting for confounders, children born to mothers had lower scores on communication and fine motor skills and had poor visual acuity when compared to children of mothers who did not work in the same industry [4].

In a follow-up study of the cohort of children born to mothers working in the Salinas Valley agricultural industry, Bouchard et al. [1] reported findings regarding neurobehavioral development of children up to seven years of age. The investigators found that markers of urinary exposure during pregnancy were negatively correlated with intellectual development in seven-year-old children, although this was not true for the cohort of children immediately after birth.

An investigation looking at lung function among children of the Salinas Valley cohort reported a decrease in pulmonary function that correlated with urinary concentrations of organophosphate exposure markers. Both forced expiratory volume in 1 second and forced vital capacity decreased significantly per ten-fold increase of urinary marker levels [8]. The study analyses were adjusted for confounders such as the mother smoking during pregnancy, ambient air particulate matter $\leq 2.5\ \mu M$, mold and pets at home, proximity of the home to highways, food insecurity, and season of the year when lung function measures were taken. As is always true, lung function measures were adjusted for individual gender and height.

6.2 Effects of Glyphosate

In contrast to the acute toxicity of the organophosphate pesticides, the acute toxicity of glyphosate is quite low. Significant amounts can be readily excreted, although this may not apply to adjuvants that are used in glyphosate's application, nor address the results of continual use and the resulting cumulative exposure. Over time, a number of investigators have identified conditions that may be associated with this type of exposure, and have also explored their consequences.

One condition that has been given significant global attention is chronic renal disease of unknown etiology. While it has been reported in a number of locations, including the upper Midwest in the US [14], Brazil in South America, and China and Taiwan in Asia [9], the major findings have come from Central America in the western hemisphere, and Sri Lanka and India in Asia [9]. Chronic renal disease of unknown etiology is a condition that affects working-age adults and farmers who do not have, in other respects, known risk factors for renal disease such as hypertension, type I diabetes, or glomerular nephritis. While it may be diagnosed in early stages of development, it is often, unfortunately, diagnosed later, frequently in areas where little medical care is available. Given the health crises that this presents, it is not surprising that many investigators have been searching for an identifiable cause or multiple causes. Two factors often cited are heat stress and dehydration (primarily but not exclusively among those working in Central American environments) [10, 11, 14]; and agricultural activity, including exposures to glyphosate [12].

A review article recently published by Redmon and associates [9] looks at peer-reviewed studies examining this condition and potential factors leading to its development in the two areas where chronic renal disease of unknown etiology is most prevalent. Their findings identified four factors that were most commonly cited (by ranking): heat stress; quality of drinking water; agrochemical products; and heavy metals. However, differences in study design, analyses of study samples, as well as the specific focus on risk factors considered in the various studies make it challenging to more definitively analyze specific causes. In some cases, a risk factor could be present in two sources (e.g., the use of agricultural chemicals as well as drinking water contamination from the same chemicals). Nevertheless, as an overall review, the article brings up a number of important points, including the need for more universal study designs, coordination between differing geographic locales, and a consideration that the condition could result from a number of factors.

While the Redmon review indirectly addresses glyphosate, a number of other sources have considered this as a more focused question. Gunatilake and coauthors [3] suggest that glyphosate may be a synergistic factor in the development of the kidney condition. A study from Sri Lanka [5] found that farmers who drank water from abandoned wells (contaminated both with glyphosate and with high levels of several heavy metals) were more likely to present with the kidney condition. While not explicitly stated, the findings implied that glyphosate, as a chelating agent, could aid in capturing the heavy metals, several of which are considered toxic. This belief, based on data available from Sri Lanka, was also cited by Wimalawansa [15].

Other investigators have considered the effects that glyphosate may play in altering or affecting bacterial species, and how the potential effects on bacteria might affect individual health. Van Bruggen and co-authors [13], citing glyphosate's effects on biosynthetic pathways in plants and bacteria, perceive the potential for selection of resistant species of both plants and bacteria. While the authors note the need for more definitive data, they also suggest that this type of pressure could lead to "distortions in microbial communities" with the potential of more antibiotic resistance as well as changes in the individual microbiome make-up. A review article by Peillex and Pelletier [6] that examines a number of chronic responses, both in vivo and in vitro that are found in human and other animal species, speculates that chronic exposure to glyphosate may result in altered immune responses. However, in contrast to these studies, Pöppe and co-investigators [7], examining clinical isolates from cattle exposed to glyphosate found that while *Salmonella* in the isolates was glyphosate-resistant, the *Salmonella* in no other ways was antibiotic-resistant. Despite this finding, the authors note that.

> the herbicide may come into contact with various microorganisms, on which it acts as an antimicrobial, and it may select for resistance and cross-resistance to clinically important antibiotics. It is therefore important to estimate whether the widespread use of pesticides may be an underappreciated source of antibiotic-resistant microorganisms that may compromise efficiency of antibiotic treatments in humans and animals.

References

1. Bouchard MF, Chevrier J, Harley KG, Kogut K, Vedar M, Calderon N, Trujillo C, Johnson C, Bradman A, Barr DB, Eskenazi B (2011) Prenatal exposure to organophosphate pesticides and IQ in 7-year-old children. Environ Health Perspect 119(8):1189–1195.DOI: https://doi.org/10.1289/ehp.1003185
2. Eskenazi B, Marks A, Bradman A, Harley K, et al (2007) Organophosphate pesticide exposure and neurodevelopment in young Mexican-American children. Environ Health Perspect 115(5):792–798. DOI: https://doi.org/10.1289/ehp.9828
3. Gunatilake S, Seneff S, Orlando L (2019) Glyphosate's synergistic toxicity in combination with other factors as a cause of chronic kidney disease of unknown origin. Int J Environ Res Public Health 16(15). DOI: https://doi.org/10.3390/ijerph16152734
4. Handal AJ, Harlow S, Breilh J, Lozoff B (2008) Occupational exposure to pesticides during pregnancy and neurobehavioral development of infants and toddlers. Epidemiology 19(6):851–859. DOI: https://doi.org/10.1097/EDE.0b013e318187cc5d
5. Jayasumana C, Paranagama P, Agampodi S, Wijewardane C, Gunatilake S Siribaddana S (2015) Drinking well water and occupational exposure to Herbicides is associated with chronic kidney disease, in Padavi-aSripura, Sri Lanka. Environ Health 14(6). https://doi.org/10.1186/1476-069X-14-6
6. Peillex C, Pelletier M (2020) The impact and toxicity of glyphosate and glyphosate-based herbicides on health and immunity. J Immunotoxicol 17(1). DOI: https://doi.org/10.1080/1547691X.2020.1804492
7. Pöppe J, Bote K, Ramesh A, Murugaiyan J, et al. (2020) Selection for resistance to a glyphosate-containing herbicide in Salmonella enterica does not result in a sustained activation of the tolerance response or increased cross-tolerance and cross-resistance to clinically important antibiotics. Appl Environ Microbiol 86(24):e01204-e1220. https://doi.org/10.1128/AEM.01204-20
8. Raanan R, Balmes J, Harley KG, Gunier RB, Magzamem S, Bradman A, Eskenazi B (2016) Decreased lung function in 7-year-old children with early-life organophosphate exposure. Thorax 71:148–153. DOI: https://doi.org/10.1136/thoraxjnl-2014-206622
9. Redmon JH, Levine K, Lebov J, Harrington J, Kondash AJ (2021) A comparative review: Chronic kidney disease of unknown etiology (CKDu) research conducted in Latin America versus Asia. Environ Res 192:1–11. DOI: https://doi.org/10.1016/j.envres.2020.110270
10. Roncal-Jimenez CA, García-Trabanino R, Wesseling C, Johnson RJ (2016) Mesoamerican nephropathy or global warming nephropathy? Blood Purif 41(1–3):135–138. DOI: https://doi.org/10.1159/000441265

11. Sabtic UP, Zanetta D, Terra M, Burdmann EA (2015) Burnt sugarcame harvesting is associated with acute renal dysfunction. Kidney Int 87(4):792–799. DOI: 10.1038/ki.2014.306
12. Seneff S, Orlando L (2018) Is Glyphosate a key factor in mesoamerican nephropathy? J Environ Anal Toxicol 8(1). DOI: 10.4172/2161-0525.1000542
13. Van Bruggen AHC, He M, Shin K, Mai V, Jeong KC, Finckh MR, Morris JG Jr (2018) Environmental and health effects of the herbicide glyphosate. Sci Total Environ 616–617:255–268. https://doi.org/10.1016/j.scitotenv.2017.10.309 Epub 2017 Nov 5
14. Wilke RA, Qamar M, Lupu RA, Gu S, Zhao J (2019) Chronic kidney disease in agricultural communities. Am J Med 132(19):e727–e732. DOI: https://doi.org/10.1016/j.amjmed.2019.03.036
15. Wimalawansa SJ (2016) The role of ions, heavy metals, fluoride, and agrochemicals: critical evaluation of potential aetiological factors of chronic kidney disease of multifactorial origin (CKDmfo/CKDu) and recommendations for its eradication. Environ Geochem Health 38(3):639–678. DOI: https://doi.org/10.1007/s10653-015-9768-y
16. Young JG, Eskenazi B, Gladstone EA, Bradman A, Pedersen L, Johnson C, Barr DB, Furlong CE, Holland NT (2005) Association between in utero organophosphate pesticide exposure and abnormal reflexes in neonates. Neurotoxicology 26(2):199–209. DOI: https://doi.org/10.1016/j.neuro.2004.10.004

Part II
The International Agency for Research on Cancer (IARC) Review Process

Chapter 7
Overview of IARC

Abstract The overall mission of IARC is "cancer research for cancer prevention" [7]. To carry out this mission, IARC brings groups of laboratory scientists, epidemiologists, and statisticians together to identify potential cancer-causing agents. The goal is to ultimately lessen the global burden of cancer-associated disease. Based on the assumption that most cancers are caused by environmental factors, IARC works to specifically identify substances or lifestyle factors that may lead to the development of cancer. The agency, nevertheless, recognizes that such factors may interact with genetic risk characteristics that increase the likelihood of disease. In carrying out its mission IARC strives to develop collaborations "across countries and organizations" [7]. A particular emphasis is on low- and middle-income countries; hence IARC seeks to include researchers from these countries. Governments of 24 countries provide funding for the overall work of the agency. Although IARC's activities are complex and many, a major endeavor is the development of monographs that examine and classify agents that may lead to the development of cancer. Primary funding for the monograph program is the US National Cancer Institute (NCI) which has provided renewable grants. Other major sources are the US National Institute of Environmental Health Sciences (NIEHS), the US Environmental Protection Agency (EPA), and the European Union (EU) Directorate of General Employment. The monograph process, which has evolved over time, has been both strongly supported, while, at times, criticized. The following chapter considers all of these issues.

7.1 Overview of IARC Activities

A significant part of IARC's work is developing monographs on agents that may be cancer hazards. While the monograph publications are a fundamental element of IARC's work, they are not the only agency activity. Other key activities include the following:

M. Richmond, *Cancer Hazards: Parathion, Malathion, Diazinon, Tetrachlorvinphos and Glyphosate*, AESS Interdisciplinary Environmental Studies and Sciences Series,
https://doi.org/10.1007/978-3-030-81953-8_7

1. A global reference for cancer registries. Key aspects of this work are:
   - The publication of "Cancer Incidence in Five Continents"
   - GLOBOCAN: an annual publication that provides estimates of cancer incidence, prevalence, mortality, and geographic variability across 20 world regions. Data provide information for 185 countries on 36 types of cancer. All cancers are broken down according to gender and age groupings
   - Classification of human tumors in the WHO "Blue Book" series
2. The establishment of a global cancer bio-bank that provides samples to researchers worldwide
3. Global education and training of cancer researchers
4. Close coordination with WHO to develop strategies for "timely policies" to control cancer.

7.2 The Process of Monograph Development

Steps in the development and publication of monographs can be summarized as follows:

7.2.1 Selection

If a substance or group of substances is to undergo review, it (they) must satisfy two criteria:

- Evidence of human exposure
- Evidence or suspicion of carcinogenicity.

Nominations for selection can come from various sources: regulatory agencies, public health organizations and officials; environmental and occupation health experts; scientists, and, since 2003, from representatives of the lay public. Beginning in 1984, an ad hoc international interdisciplinary advisory group has met to evaluate published or government-provided information and make recommendations regarding agents for review. The advisory group members have expertise in research on public health and carcinogenicity. After reviewing information on hazard potential and exposure, group members prepare a report that has recommendations for evaluation which is issued approximately 5 years before the expert group meetings to develop a given monograph are held. Ad hoc advisory groups have been convened every five years since 1984. Based on the advisory group deliberations, substances are ranked as low, medium, or high priority for evaluation.

While substances to undergo review are generally selected on the basis of recommendations by the ad hoc panels, they may also be selected subsequently based on newly developing evidence or a particular concern for public health, even if previous evaluations have led to no finding.

7.2.2 Evaluation

Approximately one year before the evaluation meetings are held, IARC selects agents for review. Following review, the findings would be published in a new volume of the monographs. Generally, agents selected for review come from the report issued by the ad hoc advisory group. Included in this selection are a preliminary list of agents, as well as a solicitation for data and experts, and a request for observers. Once agents designated for review are selected, evaluations are conducted by a Monograph meeting that involves the participation of or observation by several groups:

- A Working Group comprised of members selected for their expertise on the specific issues to be considered. The group members are charged with collecting and assessing all appropriate data on the subject(s) of interest (as designated above); for preparing summaries of these data; and for making recommendations about classification. Working Group members must be free from any conflict of interest.
- Invited specialists who also have expertise in the substance(s) under consideration, but who may have conflicts of interest. Such specialists may, however, provide their knowledge and expertise in meeting discussions.
- Representatives from national and international health agencies concerned with the substance(s) under consideration.
- Outside observers who have appropriate credentials and interests. Outside observers cannot contribute to or attempt to influence the discussions.
- Scientists designated by the IARC Secretariat. These individuals are rapporteurs or serve to facilitate the development of final analyses and reports. They can also participate in discussions.

The diversity and inclusiveness of representatives who bring a variety of backgrounds and interests are intended to ensure an open, inclusive, and thorough discussion and final assessment of any substance or substances under review.

A more detailed description of the nomination and review process can be found in a recently published article by Pearce and coworkers [11], as well as the 2019 IARC Monograph Preamble [7].

7.2.3 Classification

Starting in 1972 when the first monograph was published, IARC has evaluated a wide variety of substances that might pose carcinogenic hazards. Although the overall focus has remained essentially the same, the specific classification scheme has evolved over time. Since 1988, the IARC system ranks substances in the following order:

- Group 1: Carcinogenic to Humans: sufficient evidence of human carcinogenic potential.
- Group 2A: Probably Carcinogenic to Humans: limited evidence in humans, and sufficient evidence in animals.
- Group 2B: Possibly Carcinogenic to Humans: limited evidence in humans and insufficient evidence in animals.
- Group 3: Not Classifiable as Carcinogenic to Humans: inadequate evidence in both humans and animals.
- Group 4: Probably not Carcinogenic to Humans: evidence of human response over a large range of exposure indicating no carcinogenic response.

From 1972 to the present time, IARC has published a total of 128 monographs, classifying a total of 1,022 chemicals or other agents as known, probable, possible carcinogens, or having insufficient evidence as carcinogens. Until 2019, chemicals in category 4 (no evidence of carcinogenicity) were also listed. Table 7.1 provides a summary of the present system of classifications other than Group 4, which, as noted, was discontinued in 2019 [7].

Table 7.1 Summary of IARC monograph classifications

Category/classification	Number of substances in classification
Category 1	121
Category 2A	89
Category 2B	315
Category 3	497

7.2.4 Scope

From 1972 to 1978, the monograph series was entitled, "IARC Monographs on the Evaluation of Carcinogenic Risks of Chemicals to Man." This title was changed to "IARC Monographs on the Evaluation of Carcinogenic Risks of Chemicals to Humans" in 1978. A more substantive change was made in 1988 when the title of the monograph series became "IARC Monographs on the Evaluation of the Carcinogenic Risk to Humans." In 2019 the title became "Identification of Carcinogenic Hazards." Each of these changes has impacted the scope of the monograph series, both in terms of focus and in terms of breadth.

Reflecting the increasing breadth and depth of IARC's evaluations, the scope of monograph considerations and preparations has become far-reaching. Included today are the evaluations of a number of different chemicals, chemical mixtures, biological agents, as well as lifestyle or occupational exposures.

Below are examples of the types of substances or topical areas that have been considered:

- Chemical substances. *Examples*: specific chemical compounds used in a variety of applications (hydrochloric acid; sulfur dioxide; medications such as ampicillin; insecticides and herbicides such as lindane; chemical mixtures, such as crude oil or various fuels).
- Hormones. *Examples*: contraceptives; progestin.
- Disease-causing agents. *Examples*: Hepatitis D; HIV virus; polyomavirus; human T-cell lymphotropic virus; Schistosoma.
- Chemical implants
- Electric and magnetic fields, including low-frequency magnetic fields.
- Lifestyle factors. *Examples*: coffee and tea consumption; caffeine in coffee; ethyl alcohol in alcoholic beverages; UV tanning devices.
- Industrial processes and occupational exposures. *Examples*: petroleum refining; paint manufacture; manufacture of art glass; coal dust exposure; high temperature frying emissions.

When newer findings become available, a substance previously reviewed by IARC may be re-examined. This might lead to a revision of findings, either designating a substance as falling into a category of greater hazard or presenting a smaller hazard. For example, findings regarding lindane, a chlorinated hydrocarbon pesticide, have been published in five monographs: Volume 5, 1974 (no firm conclusions); Supplement 7, 1979b [4] (no firm conclusions); Volume 20, 1987 (category 2B); and, most recently, Volume 113, 2018 [6]. In the most recent evaluation, lindane was classified as a Group 1 agent, carcinogenic to humans. In contrast, a re-evaluation of the carcinogenic potential of coffee lowered its hazard classification. In 1991, IARC's assessment of coffee placed it in Group 2B, possibly carcinogenic. In a more recent evaluation, the IARC Working Group compared several beverages that can be consumed at very high temperatures and concluded that the temperature of the beverage, not the beverage itself was the factor determining whether or not the beverage was carcinogenic. Specifically, beverages consumed at temperatures exceeding 65 °C could be classified as probably carcinogenic, Category 2A; however, at lower temperatures, little evidence existed for carcinogenicity. However, it should also be noted that no substance classified as a Category 1 carcinogen has later been assigned to another category [14].

7.3 Evolution of the Monograph Development Process

7.3.1 Initial Establishment of the Monograph Series

In recognition of the disease burden of cancer, IARC was formed in 1965. While aspects of the agency's work have changed over time, the central mission to address the global burden of cancer and identify factors that contribute to this burden remains the same. Realizing that environmental factors play a critical role in carcinogenesis, a monograph series focusing on specific agents was proposed early in IARC's formation. As described by Tomatis recording the early work of IARC "the greatest potential for cancer control lies in preventive measures" [15]; however, how this could best be addressed required some discussion. Ultimately the agency concluded that the best way to address this issue was through a monograph series. In 1970, the IARC Advisory Committee on Environmental Carcinogenesis recommended: "… that a compendium on carcinogenic chemicals be prepared by experts. The biological activity and evaluation of practical importance to public health should be referenced and documented." The first monograph, entitled "Some Inorganic Substances, Chlorinated Hydrocarbons, Aromatic Amines, N-Nitroso Compounds, and Natural Products," was published in 1972. Clearly, the initial monograph did not specifically focus on a particular class of chemicals. In addition, while the monograph provided information about the animal and human carcinogenicity and exposure information, it did not specifically classify substances into the presently used categories. In its introduction to the initial monograph, IARC noted that:

> As new data on chemicals for which monographs have already been written and new principles for evaluation become available, re-evaluation will be made at future meetings, and revised monographs will be published as necessary [2].

7.3.2 Evolution of the Classification Criteria

Initial monographs did not specifically place evaluated substances into the later categories. Specific classifications, while considered in the early years, evolved [6]. Over time, monographs placed evaluated substances into one of the following categories: (1) sufficient evidence, indicating a causal relationship between the exposure and human cancer. (2) limited evidence, indicating a credible causal interpretation, "but that alternative explanations, such as chance, bias or confounding, could not adequately be excluded"; (3) inadequate evidence "which applies to both positive and negative evidence, indicates that one of two conditions prevailed: (a) there are few pertinent data; or (b) the available studies, while showing evidence of association, do not exclude chance, bias or confounding;" (4) no evidence from an evaluation of human epidemiology and controlled animal studies [5, 6]. Starting in 1975, conclusions were also based on results of short-term mutagenicity tests, and a more exhaustive discussion of mutagenicity tests, including types of tests, metabolic activation, and interpretation, was provided in the preamble to the 1977 monographs [3]. Techniques for data analysis and incorporation of more refined approaches to epidemiology, including meta-analyses reported or conducted at the expert panel meetings, as well as an understanding of genetic polymorphisms and the role that they play in individual susceptibility are now part of the body of knowledge of epidemiology and considered in deliberations. Other scientific information that is now considered includes toxicokinetic data, and, at the cellular and molecular levels, functional changes. Non-genetic changes such as hormonal and immunological differences must be reviewed to ensure that they are mechanistically related to cancer development.

7.3.3 *Criteria Used in Selection for Review*

While the basic criteria used by the ad hoc groups convened to select and prioritize substances for review have remained the same, in parallel with methods that have become part of the procedures used to evaluate carcinogenic hazards, methods used by the ad hoc groups also have evolved. Most recently, germane to the selection of substances to be evaluated by the 2015 expert panel that developed the 2017 Monograph on organophosphate pesticides and glyphosate was the use of chemoinformatic techniques. This was incorporated into the selection process for the 2015 evaluation by the ad hoc committee [1]. Specific steps of this process will be described in Chap. 8 which specifically focuses on the development of Monograph 112.

7.4 Criticisms of the Review Process

The basic steps in monograph development have remained in place since the program was established in 1972. However, over time, procedural and analytic changes have been made to incorporate more rigorous evaluation techniques and to establish more transparent approaches. This includes additional measures to ensure that those taking part in the evaluations are free from conflicts of interest. It is not surprising, however, given the breadth of IARC evaluations as well as the regulatory implications and economic importance of many substances that have undergone evaluation, a number of specific classifications have been questioned and methods used in evaluations from time to time have been subject to criticism.

7.4.1 Recent Criticisms

A recent commentary [11] lists several points of criticism of the IARC evaluation process. Included are the following:

General criticisms of aspects of epidemiology.

- As a scientific discipline
- Weaknesses in design, especially in case/control studies
- Investigator bias
- Publication bias, specifically that positive findings are more likely to be published
- Findings that may have involved an insufficient latency period for cancer development.

False-positive results (and its corollary, false-negative results).

- False-positive results may occur solely by chance; generally, a false-positive result will come when there is little or no expectation of an exposure-outcome relationship and will not be corroborated by other studies.
- False-negative results: resulting from weak statistical power, misclassification, or poor follow-up (such as insufficient time).

Allegations of "discontent with the IARC process".

- Generally, such criticisms appear in lay press publications. For example, Kabat (Kabat G 2012) published a statement in Forbes Magazine critical of IARC, noting that "a number of scientists with direct experience of IARC have felt compelled to dissociate themselves from the agency's approach to evaluating carcinogenic hazards. Their critique goes to the heart of the agency's epistemology and its deliberative process."

Criticisms of given evaluations, which, according to those voicing criticism, were not supported by the scientific literature.

- In the Forbes Magazine article cited above (Kabat G 2012), the IARC classification of radiofrequency waves was seen by the author as flawed because the conclusions placed "undue emphasis on a small number of positive epidemiologic studies from a single group, when the much larger body of studies indicated no elevated risk." Classification of glyphosate as a group 2A carcinogen has similarly come under criticism by several scientists [13].

Criticisms of the composition of working groups, with a specific focus on the potential of a group member to have conducted research on a subject under consideration, thus exhibiting conflict of interest. Specifically, this has involved concerns about "careerism" among scientists who "have a vested interest in advancing their own research results in the deliberations, if only to increase their prestige and future funding opportunities" [9].

7.4.2 Responses to Criticisms

Pearce and coworkers [11] provide a detailed response to each of these points. This includes the following:

General criticisms of epidemiology.

- A totality of evidence is considered
- A topic seen as important is likely to be investigated in more detail by a large number of publications, strengthening, or lessening the strength of initial findings
- Incorporation of well-developed techniques for methodological evaluation of epidemiologic studies is essential to the process of evaluation

False-positives.

- Positive findings are generally reinforced or dismissed by rigorous statistical methods as well as reproducibility
- Positive findings can be reinforced through the use of biomarkers and other methods of verification

False-negatives.

- Easier to verify or dismiss based on the use of more rigorous statistical methods, better exposure measurement, and/or biomarkers
- Use of mechanistic studies as supportive evidence

Discontent with the IARC process.

- Specific allegations (e.g., those expressed by Kabat [8] were not published in peer-reviewed literature
- No specific examples have been published in peer-reviewed literature
- Reinforcing the response published by Pearce and coworkers (13) it was noted that an estimated 1500 individuals have participated in reviews. Of these participants, "if even a small percentage [of these scientists] were disenchanted" this should be easy to document (Pearce, 13).

Specific evaluations not supported by peer-reviewed literature.

- The selection of substances to be evaluated is based on findings published in peer-reviewed literature.
- Preponderance of classifications following a critical review of peer-reviewed and published government data does not demonstrate a strong bias toward classifications as definitely carcinogenic. Out of the 1003 substances evaluated at the time of the Pierce publication, only 120 have been classified as definitely carcinogenic.

Criticisms of the composition of working groups.

- Members are selected because of their expertise and are highly accomplished investigators.
- Generally, several agents are considered, making it unlikely that a Working Group participant would consider only one agent.
- Expertise of members can provide valuable information for the group as a whole
- Decisions are based on a consensus of all Working Group members

In a concluding statement, the authors stated the following:

> We concluded that these recent criticisms are unconvincing. The procedures employed by IARC to assemble Working Groups of scientists from the various disciplines and the techniques followed to review the literature and perform hazard assessments of various agents provide a balanced evaluation and an appropriate indication of the weight of the evidence. Some disagreement by individual scientists to some evaluations is not evidence of process failure. The review process has been modified over time and will undoubtedly be altered in the future to improve the process. Any process can, in theory, be improved, and we would support continued review and improvement of the IARC processes. This does not mean, however, that the current procedures are flawed [11].

A companion review written by Samet [14] provides a summary discussion of these same issues, and notes in concluding:

> the types of concerns raised about the IARC monograph program are also archetypical of strategies for creating 'doubt' about scientific evidence that has policy implications. Such strategies can be traced to the 'playbook' of the tobacco industry for discrediting findings related to active and passive smoking … One tactic has been to question the processes used to draw causal inferences and the integrity and potential conflicts of interest of those doing so. The IARC processes are robust and transparent and as concluded by Pearce and his 123 colleagues, not flawed and biased.

7.5 IARC's Relationship with Other Agencies

While IARC does not conduct quantitative risk assessments for regulatory purposes, its findings provide information on carcinogenicity for the WHO's publication of global burden of disease estimates. Originally developed in the 1990s as a project between the Harvard School of Public Health, the World Bank, and the WHO [10], the global burden of disease estimates parameters of health by comparing data for a population of a given age, gender, and geographic location to a reference group. Included as parameters are the disability-associated lost years (DALYs) which can be thought of "as one lost year of 'healthy' life and the measured disease burden" based on comparison with a standard population. DALYs are essentially the sum of two measures: the years of life lost and the years of life with a disability for a given population, based on age group, gender, and time (WHO, 2017). As new information has become available, estimates of the global burden of disease have been revised. Each covers a given period and includes projections, given the available information at a specific time.

It is significant that not all IARC findings can contribute equally to the global burden of disease estimates. With respect to pesticides and herbicides, Prüss-Ustün and coworkers [12] in a review of "knowns and unknowns" note that incomplete exposure information limits assessment of the burden of disease from chronic exposure to pesticides and herbicides.

IARC is a key player in assessing the carcinogenic potential of specific agents, but, necessarily, not the only agency addressing carcinogenicity. Among organizations in the United States, the Environmental Protection Agency (EPA) and the National Toxicology Program (NTP) both assess carcinogenic potential. The EPA also has regulatory authority and, therefore, conducts quantitative risk assessment as part of its mission. In the European Union, overall regulation (which incorporates hazard identification) of chemical substances is under the Registration, Evaluation, Authorization, and Restriction of Chemicals (REACH) regulation which was placed into practice in 2007. REACH covers a number of different classes of substances, ranging from chemical substances to biocides, and to air pollutants. While the specifics of a given regulation may differ according to the substance type, the initial step requires hazard identification. Since the regulatory process also considers exposure, however, REACH may determine that a substance identified as posing a carcinogenic hazard is unlikely to be carcinogenic for a given population. Globally, recognition of hazard identification and regulation is quite variable; to some extent, it is more effective in developed or developing nations, although it is difficult to generalize.

References

1. Guha N, Guyton KZ, Loomis D, Barupal DK (2016) Proritizing chemicals for risk assessment using chemoinformatics: examples from the IARC monographs on pesticides. Env Health Perspect 124(12):1823–1829. https://doi.org/10.1289/EHP186
2. IARC (1972) "IARC monographs on the evaluations of the carcinogenic risk of chemicals to man. Some Inorganic Substances, Chlorinated Hydrocarbons, Aromatic Amines, N-Nitroso Compounds, and Natural Products." vol. 1
3. IARC (1977) "IARC monographs on the evaluations of the carcinogenic risk of chemicals to man. Some Fumigants, the Herbicides 2,4-D and 2,4,5-T, Chlorinated Dibenzodioxins and Miscellaneous Industrial Chemicals." vol. 15
4. IARC (1979) "IARC Monographs on the evaluations of carcinogenic risks to humans. Overall evaluations of carcinogenicity: an updating of IARC monographs, vol 1–42." IARC Monographs, supplement 7
5. Athouse R, HuffJ, Tomatis L, Wilbourn J (1980) An evaluation of chemicals and industrial processes associated with cancer in human and animal data: IARC Monographs 1 to 20. Cancer Research 40(1):1–12
6. IARC (2018) "IARC Monographs on the evaluations of carcinogenic risks to humans. DDT, lindane, and 2,4-D." vol. 113
7. IARC (2019) "IARC Monographs on the identification of carcinogenic hazards to humans: preamble." https://monographs.iarc.fr/wp-content/uploads/2019/01/Preamble-2019.pdf
8. Kabat G (2012) "How activism distorts the assessment of health risks." Forbes, Nov 20
9. McLaughlin JK, Bofetta P, LaVecchia C, Lipworth L et al (2010) Authors' response: problems with IARC's expert working groups [letter]. Int J Epidemiol 40(6):1728–1729. https://doi.org/10.1093/ije/dyr158

10. Murray CJ (ed) (1996) The global burden of disease: a comprehensive assessment of mortality and disability from diseases, injuries, and risk factors in 1990 and projected to 2020. World Health Organization. https://apps.who.int/iris/handle/10665/41864
11. Pearce N, Blair A, Vineis, P, Ahrens W et al (2015) IARC monographs: 40 Years of evaluating carcinogenic hazards to humans. Environ Health Perspect 123(6):507–514. https://doi.org/10.1289/ehp.1409149
12. Prüss-Ustün A, Vickers C1, Haefliger P, Bertollini R (2011) Knowns and unknowns on burden of disease due to chemicals: a systematic review. Environ Health 10(9). https://doi.org/10.1186/1476-069X-10-9
13. Richmond ME (2018) Glyphosate: a review of its global use, environmental impact, and potential health effects on humans and other species. J Environ Stud Sci 8(4):416–434. https://doi.org/10.1007/s13412-018-0517-2
14. Samet J (2015) The IARC monographs: critics and controversy. Carcinogenesis 36(7):707–709. https://doi.org/10.1093/carcin/bgv062
15. Tomatis L (1975) The IARC Program on the evaluation of the carcinogenic risk of chemicals to man. Ann NY Acad Sci 271:396–409. https://doi.org/10.1111/j.1749-6632.1976.tb23139.x

Chapter 8
Development of IARC Monograph 112: Assessment Process, Review Panel Decisions, and Basis for Final Panel Decisions

Abstract Prior to its 2015 evaluation, IARC had considered the four organophosphate pesticides and glyphosate as candidates for assessment. These were either assessed and given low carcinogenicity rankings or, in the case of glyphosate, assigned a low priority and not evaluated. However, the development of new information and new techniques, both for consideration of substances to be evaluated and for the assessments themselves, led to a 2014 IARC decision that parathion, malathion, diazinon, tetrachlorvinphos, as well as glyphosate, should be evaluated as a group. A complete description of the selection process and factors that led to this decision is published in IARC Monograph 112, "Some Organophosphate Insecticides and Herbicides." Specific details of the evaluations of each compound are also published in Monograph 112. The following chapter includes a discussion of updated selection techniques, issues considered in the evaluation of each compound, evaluation findings, a critique of findings, and the final decisions by the review panels.

8.1 Selection, Exposure Assessment, and Evaluation Techniques

8.1.1 Selection

As noted in Chap. 7, IARC has always had two general criteria for selecting agents for review, namely evidence of human exposure and some evidence or suspicion of carcinogenicity [16]. While these general criteria have been applied since the Monograph series was first established, specific techniques for selection have become more detailed and sophisticated as technology has developed.

In a recent publication, Guha and coworkers [15] provide details of the selection process for the two monographs that were developed in 2015. With respect to the organophosphate pesticides and glyphosate, the IARC Advisory Group assembled prior to the most recent evaluations recommended that the program assess pesticides, a process that led to an initial identification of roughly 6,000 "relevant chemical

M. Richmond, *Cancer Hazards: Parathion, Malathion, Diazinon, Tetrachlorvinphos and Glyphosate*, AESS Interdisciplinary Environmental Studies and Sciences Series,
https://doi.org/10.1007/978-3-030-81953-8_8

structures." It was briefly noted in Chap. 7 that the Advisory Group used a bioinformatics approach to identify specific pesticides for evaluation. Using this approach, the substances were organized into network maps based on chemical similarity. Individual pesticides, assigned nodes in the network maps, were linked to other nodes if their chemical similarity was > 60%. At the same time, an exhaustive automated literature search was conducted using PubMed databases. The data from both searches were integrated to identify pesticides for possible evaluation. Findings indicated that "top ranked" pesticides fell into two main groups: organophosphate and organochlorine compounds. As a final step in the selection process, other criteria, including volume of use, IARC advisory group priority, prior classification if applicable, and other classifications (e.g., EPA classification) were considered. With respect to glyphosate, which has some structural similarities to organophosphate pesticides, an important criterion in selection was its widespread use, even though it was given a lower ranking using the chemoinformatics approach. It is noteworthy that, even though a growing volume of publications related to specific areas of cancer mechanisms exists, mechanistic studies were not incorporated into the bioinformatics approach.

8.1.2 Exposure Assessment

Key to evaluating epidemiology studies are accurate exposure data. However, as noted in the narrative of Monograph 112, accurate assessment of pesticide exposures is challenging for a number of reasons, including the following:

- Lack of a good bio-marker that is specific to a given pesticide
- Route of exposure which may influence intake
- Sensitive measures of exposure
- Intensity of application, including spray pressure (as applicable) to measure exposure intensity and a timeframe
- Multiple exposures to a number of substances, which can be either closely related chemically (such as organophosphate pesticides), or of different compound classes
- Individual employment history, including types of farms and farming industries where an agricultural worker has been employed. Such histories can provide important information about the possible number of past exposures as well as the likelihood of farm workers using personal protective equipment
- Sampling techniques
- Use of protective clothing
- Influence of climate that can, among other things, determine the use of personal protective equipment.

Although short-term biomarkers found in urine and blood may be specific to a given pesticide, these are excreted within two to three days following exposure and provide no long-term data. Furthermore, when a pesticide belongs to a group of chemically related compounds, as is true with the organophosphate pesticides, it is difficult to obtain detailed compound-specific information.

Some of the more comprehensive information has come from the Agricultural Health Study (AHS), a long-term cohort study examining outcomes for a number of chemicals used in farming [1]. Initiated in the 1990s and sponsored by several US government agencies including the National Cancer Institute (NCI), EPA, and the National Institute of Environmental Health Sciences (NIEHS), the AHS has developed individual information about long-term exposures and health outcomes for a number of substances, including the organophosphate pesticides and, over a shorter timeframe, glyphosate in two states with extensive farming, North Carolina and Iowa. Those enrolled in the study include farmers and commercial workers who are licensed or certified pesticide applicators as well as their spouses and children. During the first years of the study, approximately 27,000 subjects were enrolled [1]. In the years following, a total of approximately 57,000 subjects have been enrolled.

As noted by Alavanja et al. [1], the study has several objectives, namely to:

(1) identify and quantify cancer risks among men, women, whites, and minorities associated with direct exposure to pesticides and other agricultural agents;
(2) evaluate non-cancer health risks including neurotoxicity, reproductive effects, immunologic effects, non-malignant respiratory disease, kidney disease, and growth and development among children;
(3) evaluate disease risks among spouses and children of farmers that may arise from direct contact with pesticides and agricultural chemicals used in the home lawns and gardens, and from indirect contacts, such as spray drift, laundering work clothes, or contaminated food or water;
(4) assess current and past occupational and non-occupational agricultural exposures using periodic interviews and environmental and biologic monitoring;
(5) study the relationship between agricultural exposures, biomarkers of exposure, biological effect, and genetic susceptibility factors relevant to carcinogenesis;
(6) identify and quantify cancer and other disease risks associated with lifestyle factors such as diet, cooking practices, physical activity, smoking and alcohol consumption, and hair dye use.

The AHS has been divided into four phases. The first, from 1993 to 1997 involved subject recruitment and enrollment as well as a detailed pesticide history for applicators enrolled in the study; the second, from 1999 to 2003 involved a five-year follow-up phone interview, collection of buccal cell samples, updating pesticide use, and a questionnaire regarding diet; the third phase, from 2005 to 2010 involved another phone interview and update regarding pesticide use as well as an update of medical history; the fourth phase, from 2013 to 2015 involved an update of medical history.

A distinct advantage of the AHS is that it is a long-term cohort study which addresses concerns about recall bias inherent in case–control studies. Given the generally long latency period from exposure to the development of cancer, this is especially advantageous. However, obtaining accurate exposure histories for pesticide work is challenging for the reasons cited above. In an attempt to address many of these issues, the AHS developed an exposure algorithm incorporating four items: (1) exposure from mixing; (2) exposure from applying; (3) exposure from repairing pesticide application equipment; (4) use of personal protective equipment, PPE. In its original form, the algorithm had the following formula:

$$\text{Exposure Intensity Score} = ([\text{MIX}] + [\text{APPLY}] + [\text{REPAIR}]) \times [\text{PPE}])$$

Use of the algorithm taken together with detailed questionnaire information could, in theory, provide accurate exposure information [6].

Clearly, over time, information regarding pesticide use may change for individual applicators. As noted in Monograph 112 [16], some applicators may work at large industrial farms producing one or two crops and are likely to be exposed to a "handful" of substances over a working lifetime; others working in facilities involved in such activities as culture of flowers and a variety of seasonal vegetables can be exposed to as many as 10–15 substances per farming season. The route of exposure is also an important consideration. Although pesticide applications are generally through some type of spraying, it is generally believed that the primary route of exposure is dermal, and overall, including applicators as well as others who work in fields, hands are the most likely site of exposure unless a worker wears good protective gloves. To assess dermal exposure, a variety of sampling strategies have been used, including the use of dermal patches or other external clothing such as shirts to estimate intake. Dermal exposure data can be combined with information on the time a worker has spent spraying, application spray pressure, and climactic conditions to establish an overall exposure profile. It is noteworthy that air sampling has generally yielded poor data, and, with respect to spouses and children of applicators who have provided samples for short-term biomarkers, the biomarkers are often negatively correlated with air sampling data [16]. Furthermore, while intensity algorithms correlate well with actual measurements using personal exposure data, correlations between urinary samples were moderate, and intensity algorithms correlate poorly when using air exposure data.

In conclusion, when looking at epidemiologic studies, IARC the Monograph noted that.

> working group members considered strengths and limitations of the exposure assessment methods used in each study and took these into account in its evaluations

8.1.3 Genotoxic Evaluation

Part of the analysis of the in vitro bioactivity of the pesticides reviewed in IARC Monograph Volume 112 used data from high-throughput screening assays generated by the Toxicity Testing in the twenty-first century (Tox21) and Toxicity Forecaster (ToxCast™) research programs developed by several US government groups: the US EPA, the NIEHS together with the National Toxicology Program (NTP), and the National Human Genome Research Institute/National Institutes of Health Chemical

Genomics Center [29] The high-throughput screening assays Tox21 and ToxCast ™ allow multiple assays across a range of concentrations and endpoint groupings of bioactivity [28]. In 2014, the Advisory Group to the IARC Monographs program encouraged the use of this process to evaluate pesticides. The Working Group members, as well as IARC staff, developed information about malathion, diazinon, parathion, and tetrachlorvinphos as well as the metabolites malaoxon, diazoxon, and, to a lesser extent, paraoxon, looking at available data on 10 characteristics of known human carcinogens as designated by Smith and co-authors in 2016 [27]. The ten key characteristics are whether the substance is:

1. Electrophilic or can be metabolically activated
2. Genotoxic
3. Alters DNA repair or genomic instability
4. Induces epigenetic alterations
5. Induces oxidative stress
6. Induces chronic inflammation
7. Immunosuppressive
8. Modulates receptor-mediated effects
9. Can cause immortalization of some cell types
10. Alters cell proliferation, cell death, or nutrient supply.

IARC's inquiry found that the pesticides and their metabolites as a group had positive responses to seven of the designated 10 characteristics. It is noteworthy that an expert Working Group convened by IARC in 2012 concluded that carcinogens in Group 1 (carcinogenic to humans) commonly show one or more of these 10 key characteristics.

In using the high-throughput system, once Tox21 and ToxCast™ information was gathered, the cumulative score of a given substance was given a ToxPi score, essentially a visual representation as to the size and radius of a pie slice for a given characteristic. More specific details of this inquiry can be found in Monograph 112. While it was possible to use this approach for evaluating the four pesticides and some of their metabolites, data for glyphosate were not available.

In addition to the high-throughput system used for the four pesticides, genotoxicity data for cells of tissues taken from humans and other animals exposed in vivo as well as human and other cell samples exposed in vitro to the pesticides and/or their metabolites and to glyphosate and/or its metabolites were assessed for significant changes. A number of genotoxicity endpoints were used in inquires. These are described in Table 8.1.

Table 8.1 Endpoints used in genotoxic assays

Toxicity test	Description	Remarks
Sister chromatid exchange	The ability of a chemical to augment the exchange of DNA between two sister chromatids during chromosomal duplication	
Micronucleus formation	Chemical's ability to cause chromosomal damage as illustrated by the presence of micronuclei	Micronuclei can contain chromosomal fragments or whole chromosomes from disruption of the mitotic apparatus
DNA comet assay	Individual cell DNA damage assay. Fragments from damaged DNA, unlike intact DNA, can migrate outside a cell, while intact DNA remains attached to cell membrane during agarose electrophoresis. The undamaged DNA is the comet "head" and the damaged DNA is the comet tail, as seen by staining	Size of comet tail can indicate the extent of DNA damage
Chromosomal aberrations	Missing or irregular chromosomes, including detachments and re-attachments. When used in a genotoxicity test, if a substance is genotoxic, the number of aberrations should increase with an increasing dose of the substance being tested	
Dominant lethal mutation assay	Assay for a substance that has induced a lethal change in DNA that will not be lethal to the gamete, but will be lethal for the developing fertilized egg or embryo	
HPRT assay	HPRT (hypoxanthine phosphoribosyl transferase) gene is found on the X chromosome of mammalian cells. The toxicity of a given chemical can be tested by administering increasing amounts of chemicals to mammalian cells	Cells are given a toxic analog to a normal HPRT substrate. Mutant cells, unable to metabolize the analog are detected by comparison to non-mutant cells
Reverse mutation	Generally, a reversion of a mutant gene to wild type gene in a biosynthetic pathway	Used in prokaryotic cells to determine genetic reversion

8.2 Evaluations of Individual Pesticides

A detailed summary and brief critique of the evaluation of each substance are provided in the following sections. The discussion regarding studies used in the evaluations is intended to provide a synopsis of the findings, rather than to provide specific details. More specific information can be found in the tables included in Monograph 112. The summaries and critiques of the pesticides are provided in the same order as they were presented in the IARC Monograph. Information on glyphosate is presented following the pesticide summaries.

8.2.1 Malathion

8.2.1.1 Epidemiology Evaluations

Including publications from the AHS, a number of epidemiology studies have been conducted to evaluate malathion's long-term toxicity. The 2015 monograph lists two cohort studies, including the Agricultural Health Study (AHS) current to 2015 [19], and a study conducted in Italy, published in 1999 [25]; two nested case–control studies; and 12 case–control studies [16]. A small majority of the case–control studies (7) present findings in the US and Canada; the remaining five studies were conducted in EU member countries. Other than a few studies examining pesticide effects on pregnancy outcomes, very few studies have examined specific effects on women [16].

A closer examination of each study provides a more in-depth "rating" of those deemed most relevant to the IARC Working Group evaluation. Perhaps the most detailed of the studies, namely the AHS, has over time enrolled and obtained data from 52,395 applicators and 32,347 spouses and their children. The study has also looked at a number of cancer endpoints, rather than focusing on a few types of cancer. And, as noted earlier, because it is a cohort study, in theory it addresses the concern about recall bias that may be inherent in case–control studies.

At the same time, the Working Group also identified case–control studies that were seen as having strengths. The Mid-West Case–Control study, which looked at a number of cancer cases and controls in the Upper Mid-West, was regarded as strong. Analysis of data included adjustments for more than one exposure, and a pooled data set made it possible to analyze for specific pesticides, allowing for increased specificity. The Cross-Canada study, which included a number of different provinces where farming practices varied, was seen as having high quality and detailed collection of information about pesticide use. This included stratification according to hours of pesticide use (<10 h vs. ≥ 10 h of use) per year. Cases and controls were matched according to such variables as location, age, family history of diseases, and others [16].

While the Florida Pest Control Worker Study [3] was considered to be a strong investigation, it was limited in that it included proxy interviews for cases that could lead to exposure misclassification [16].

The United Farm Workers study in California [3] was regarded as having both strengths and weaknesses. It did not involve self-reporting, which may effectively eliminate recall bias. Exposure information was established through a variety of other means. This included union data about crops and the types of pesticides that would have been used in growing specific crops; a linkage to the California Cancer Registry (which records static information); ecological data about exposure (information from the California registration of pesticide use/county). An additional strength was seen as the lapse between the likely time(s) of exposure and outcome (approximately 20 years) [16].

While the specific focus on types of cancer varied from study to study, overall, a common thread was the development of hematopoietic or closely related cancers. One cancer endpoint that appeared frequently was non-Hodgkin's lymphoma. That a relatively rare cancer was frequently seen is noteworthy. Nevertheless, the Monograph includes discussion pointing out that, in a large cohort, especially where misclassification of exposures may occur, this occurrence could have happened by chance [16].

8.2.1.2 Controlled Exposure Animal Studies

A total of six controlled exposure long-term animal studies focusing on cancer development in animals given malathion were reviewed: two studies of $B6C3F_1$ mice treated with malathion; and two studies of Fisher 344 rats and two studies of effects in Sprague–Dawley rats. In several, the data that were examined represent an extension of an earlier study or a pathological re-examination of tissues from an earlier study. As well as malathion, the effects of malaoxon were examined in Fisher 344 rats and in $B6C3F_1$ mice. Finally, the effects of malathion on the development of the mammary gland were studied by subcutaneous injection of malathion.

Overall, while the study findings were not unequivocal, a number did show positive trends as the result of treatment with malathion. Some findings were sex-dependent. An EPA study reported in 1994 [26] found a significantly positive dose-related trend in both hepatocellular adenomas and carcinomas combined. However, a later pathology examination of these tissues suggests that more of the tissues examined were adenomas.

A study conducted by the NTP in 1978 [20] of mice reported a significant trend in follicular cell adenomas or carcinomas of the thyroid gland; however, a subsequent pathology re-analysis reported no significant trend in the development of these tumors [20]. Findings from a 1979 NTP study [21] reported a significant trend of pheochromocytoma in male Fisher 344 rats but a non-significant trend in female rats.

In 1997 EPA reported that male Fisher 344 rats given malathion were found to have a significant trend of a rare nasal tumor, while females were reported to show a significant trend in the development of hepatocellular adenomas and carcinomas combined [13].

Findings from long-term exposures studies conducted by EPA in 1984 and 1991 using Sprague–Dawley rats reported significant mammary tumors in female animals from treatment with malathion. Possibly, as a follow-up, in an investigation in which rats were given subcutaneous injections of malathion, abnormalities were found in mammary gland development. However, as noted in Monograph 112, these findings did not provide specific data for individual animals [16].

Controlled exposure studies also investigated the effects of malaoxon. Findings in mice were found to be non-significant. Male Fisher 344 rats given malaoxon developed cell hyperplasia of the thyroid gland. Female rats showed the development of C cell adenomas and carcinomas, although the statistical analysis "bordered" on non-significant. In rats, there was a dose-related decrease in survival time and an increase in mononuclear leukemia in the males. However, this condition is common in Fisher 344 rats [16].

8.2.1.3 Genotoxic Effects

Complementary findings from epidemiology studies and controlled exposure animal studies are investigations that generally look at in vitro endpoints from exposure to malathion or malaoxon. While some of these tests were conducted using prokaryotic cell systems (a total of eight), the bulk (71) are results from eukaryotic cell studies.

Monograph 112 reports results of approximately 79 such studies. This includes cultured human cells taken from humans exposed in vivo (14); in vitro studies of human cells (14); cultured cells from non-human mammals exposed in vivo (22); in vitro exposure of non-human mammal cells (10); in vivo exposures of other animals, specifically fish species and drosophila (10); one study of yeast cells; and eight studies of bacterial cells. The great majority of findings were positive and significant [16].

In addition to an assessment of publications in the literature regarding mutational tests, the high-throughput Tox21 and Toxicity Forecaster (ToxCast™) programs were used to obtain more dose-specific data about groupings of ten measures of genotoxicity. Visual representations of these tests were shown in ToxPi diagrams that can be found in Monograph 112. Positive findings are summarized in Table 8.2.

8.2.1.4 Summary of Findings

In assessing information on epidemiology, controlled animal exposure studies, and mutagenicity, the Working Group noted that there is limited evidence of carcinogenicity from epidemiological studies, with the strongest evidence being the development of non-Hodgkin's lymphoma; sufficient evidence from controlled exposure

Table 8.2 Summary of Tox21 and ToxCast™ findings of malathion

Test	Malathion ranking	Malaoxon ranking
Metabolic Activation	High Cyt 450 inhibition	Low Cyt 450 inhibition
Epigenetic Alterations	Low DNA binding	Inactive
Epigenetic Alterations	Low activity transformation catalysts	Inactive
Oxidative Stress	Overall intermediate activity	Overall intermediate activity
Chronic inflammation	Negative cytokine	High cytokine
Receptor-mediated effects	Intermediate	Low
Cytotoxicity and cell proliferation	Little or no effect	Negative

animal studies; and strong in vitro evidence of genotoxicity, in vitro evidence of receptor modulated effects related to tumor development, as well as in vivo evidence of oxidative stress in animals. The Working Group concluded that malathion should be classified as a Category 2A Probable Carcinogen.

8.2.2 *Parathion*

Parathion use has been banned, highly restricted, or not registered for use in a number of countries. California, which is one of few geographic entities in the US reporting pesticide use, noted that in 2017 it was used in only one county. Despite all of the bans, however, as noted earlier, on a global basis parathion residues on food have been reported within recent years in Ghana [7] and China [2]. Nonetheless, it is not surprising, given the paucity of information regarding its use, few recent epidemiologic studies have been published. Data regarding exposures are old, generally dating from the 1970s and 1980s, and, in addition, no specific biomarkers for parathion exist [16].

8.2.2.1 Epidemiology Evaluations

Earlier epidemiology evaluations were inconclusive. Although considered by earlier IARC, monographs published in 1983 and 1987 classified parathion in Group 3, unclassifiable as a carcinogen. More recently, parathion was studied as part of the AHS cohort investigation, and a 2005 publication of findings from the AHS investigations noted that in the Iowa cohort an increase in breast cancer was found among spouses of men who used parathion in their work [16]. While the risk, although noted, was non-significant for all women, it became significant in postmenopausal women or those with a family history of breast cancer [16]. The AHS also reported a significant increase of melanoma among workers with more than 56 days of exposure to parathion. The study was controlled for factors such as skin color, time spent in the sun, and tendency to burn. These findings may, however, be complicated by the potential effect modification of lead arsenate, also used in many of the same agricultural operations. A significantly increased incidence of prostate cancer was reported among workers with a family history of prostate cancer and who were applying high volumes of parathion [16].

Findings from a case–control study published in 2005 of non-Hodgkin's lymphoma among farm workers in Kansas, Iowa, Minnesota, and Nebraska reported non-significant findings. A case–control investigation conducted in British Columbia examined all prostate cancer cases and controls over the period 1983–1990 and used a job-exposure matrix to evaluate the potential impact of pesticide exposure. While the study results indicated some trend with low vs. high use of parathion, the IARC Working Group noted that the results should be regarded with caution given the potential likelihood of exposure misclassification in case–control studies [16].

8.2.2.2 Controlled Exposure Animal Studies

The Working Group considered the results of a study on cancers in $B6C3F_1$ mice reported in 1979 by NTP [22]. The experimental results reported no significant findings. An EPA study in 1991 detected the development of bronchoalveolar adenomas and carcinomas (especially combined) at the lowest dose level; however, animals in the lowest dose level had mistakenly been given a dose higher than the highest experimental dose level. As a result, they were taken off of any dose to allow for recovery; several died. Thus, the importance of significant findings among animals in the "lowest dose" level is unclear.

Early controlled animal exposure studies of rats were conducted in the 1950s, with no significant findings. The Working Group noted that the interpretation of these studies was limited by the small number of exposed animals in each exposure group as well as the relatively short exposure period [16]. A study using Osborne-Mendel rats reported by NTP in 1979 found trends in adrenocortex, thyroid, and pancreatic isle tumors [22].

Two studies using Sprague–Dawley rats were considered. The first, reported in 1984 [4], noted non-significant thyroid follicular cell adenomas in male rats. The second study reported in 2001 involved subcutaneous injections in female rats. Rats treated with parathion were found to a develop significant number of mammary adenocarcinomas [5].

A study of Wistar rats reported in 1989 [24, 30] noted a significant positive trend in the development of pancreatic tumors, exocrine adenoma/carcinoma combined, and islet cells in males. No findings of significance were reported in females.

8.2.2.3 Genotoxic Effects

Studies examining the genotoxic effects of parathion and its major metabolite paraoxon were reported for human cells, for cells of several other mammalian species, as well as non-mammalian species. Lymphocytes taken from humans exposed in vivo to pesticides that included parathion were found to have chromosomal aberrations and to exhibit sister chromatid exchange when cultured. Human hepatocytes exposed in vitro to parathion were found to have strand breaks and test positive in comet assays. Fetal lung fibroblasts exhibited unscheduled DNA synthesis, and lymphoid cells exposed to parathion or paraoxon exhibited DNA breaks, positive comet assays, sister chromatid exchange (parathion exposure only), and micronuclei (paraoxon exposure only).

Ovarian cells taken from female mice exposed in vivo to parathion were found to have dominant lethal mutations, and bone marrow cells were reported to have micronuclei in three separate studies. In vitro investigations of Chinese Hamster ovary cells reported positive findings for sister chromatid exchange when exposed either to parathion or paraoxon. Chinese Hamster lung cells exposed to parathion exhibited micronuclei, while rat primary hepatocytes tested negative for unscheduled DNA synthesis.

Non-mammalian systems, prokaryotic or eukaryotic, reported equivocal findings. *Salmonella typhimurium* reported negative findings for reverse mutational assays with both parathion and paraoxon, with or without activation. A drosophila assay for a sex-linked recessive lethal mutation was also negative. The remaining assay reported a positive finding in a forward mutation in a yeast strain (more details not available) [16].

Parathion was also analyzed using the Tox21 ToxCast ™ high-throughput analysis. Summary results are presented below (Table 8.3).

Table 8.3 Summary of Tox21 and ToxCast™ findings of parathion

Test	Parathion ranking	Paraoxon ranking
Metabolic activation	High Cyt 450 activation	High aromatase inbition
Epigenetic alterations	Low DNA binding	Negative DNA binding
Oxidative stress	Intermediate oxidative stress marker	Negative oxidative stress marker
Oxidative stress	Negative oxidative stress	Negative oxidative stress
Oxidative stress	Intermediate metalloproteinase	Negative metalloproteinase
Receptor-mediated effects	Intermediate to high	Low

8.2.2.4 Summary of Findings

In its overall assessment of findings, the Working Group judged the findings from epidemiology studies to be inadequate. While some findings from case–control studies indicated elevated odds ratios, overall the results "did not reach statistical significance" and "have not been replicated in separate studies" [16]. Animal studies did provide sufficient evidence for carcinogenicity. While in vitro studies, especially those with mammalian cell lines provided positive evidence, not all did. In conclusion, the Working Group classified parathion as possibly carcinogenic, Category 2B.

8.2.3 Diazinon

Originally having both household, and farming and nursery applications, personal household use is now banned. However, older epidemiology studies may include findings from personal household use.

8.2.3.1 Epidemiology Findings

Epidemiology findings considered by the Working Group come both from several cohort studies and case–control studies. Cohort studies include the Florida Pest Control Study, the United Farm Workers Study (California), and reports from the AHS. Case–control studies include the Mid-West Study, the Upper Mid-West Study, the Cross-Canada Study, and a Canadian study conducted by McDuffie and coworkers [18]. Generally, the focus of all has been on large-scale agricultural or nursery use, although, as noted, one also looked at cancers related to household use (prior to diazinon being banned for household use).

Both the Florida Pest Control Study and the United Farm Workers Study included nested case–control investigations. The nested study conducted in Florida compared the incidence of lung cancer among workers who had ever used diazinon versus those who had never used the pesticide. While point estimates in both studies were relatively high, they were not statistically significant as measured by 95% confidence intervals. The nested case–control studies in the United Farm Workers investigation examined total leukemia, lymphocytic leukemia, granulocytic leukemia, non-Hodgkin's lymphoma, and breast cancers among farm workers (as opposed to applicators). Point estimates of most studies were greater than 1.0, but none approached significance [16].

The AHS includes results from a number of studies that have been reported since AHS's inception. Findings include results from investigations among both North Carolina and Iowa farmers, as well as findings that looked only at the North Carolina or only at the Iowa cohorts. Exposures were generally classified as lifetime exposure days. Other categories included lifetime exposure days times exposure intensity, and ever versus never exposed (where multiple pesticides were used). Lung cancer was reported in the combined North Carolina and Iowa cohorts as well as in the North Carolina cohort alone. This included a significant positive trend as well as a significant finding of lung cancer among farmers in the high exposure group. A positive trend and statistically significant finding for the high exposure farmers was also seen in the North Carolina cohort taken as a separate group. However, no findings of significance were reported for non-Hodgkin's lymphoma, lung, multiple myeloma, melanoma, follicular, bladder, kidney, prostate, or colorectal cancers in a later examination of the combined cohorts [16].

The case–control Mid-West USA study produced reports from studies conducted in Iowa, Minnesota, Kansas, and Nebraska. This group of studies, one of which was conducted over a number of years, including reports of follow-up investigations in 2003, also involved investigations of a number of cancers. With the exception of a positive significant finding for the development of non-Hodgkin's lymphoma among workers using diazinon before 1965, no significant findings were reported for a variety of other cancers [11]. Another study published in 2003 reported significant findings for non-Hodgkin's lymphoma among farmers who had ever used diazinon. However, when corrected for co-exposure to atrazine, the findings were no longer significant [16].

The Upper Mid-West study, looking at cancers of the brain, stomach, and esophagus, reported no significant findings. A study reported in 1993 [8] that looked at the development of childhood brain cancer related to in utero or early development exposure to diazinon during a period when the pesticide was still approved for household use did report positive findings when using information from the child's mother.

Findings from a study looking at the incidence of multiple myeloma or Hodgkin's lymphoma noted high odds ratios for both; however, results of neither analysis were found to be significant. A Canadian study published in 2001 reported the same findings for non-Hodgkin's lymphoma. The Cross-Canada investigation reported a significant finding for soft tissue sarcoma associated with diazinon exposure [16].

One meta-analysis using data from three case–control studies reported significant results for non-Hodgkin's lymphoma [16].

8.2.3.2 Controlled Exposure Animal Studies

The Working Group considered data from three controlled exposure studies. Two were reported by NTP in 1979 [23], and the third was reported by EPA in 1993. The NTP studies involved feeding B6C3F$_1$ mice and Fisher 344 rats for a period of 103 weeks. Mice in the low dose group were found to have an increase in hepatocellular carcinoma, although this was not observed in other dose groups. Male rats in the lower dose group had a slightly higher incidence of leukemia and lymphoma combined than the historical incidence of these neoplasms in control groups. No other increases were observed. The IARC Working Group concluded that findings reported from these studies could not be related to diazinon exposure [16].

In 1993 EPA reported that in a controlled exposure study of Sprague–Dawley rats, mortality was greater in the lower dose group. No neoplasms were seen, and the study was terminated early [16].

8.2.3.3 Genotoxic Effects

In contrast to the few controlled exposure animal studies discussed by the IARC Working Group, a large number of studies examining genotoxic effects were considered. These included effects seen in cell samples taken from exposed humans; in vitro studies of human cells exposed to diazinon, or its metabolites diazoxon and diethylthiophosphate; cell samples taken from non-human mammals and other species exposed in vivo; and in vitro exposure studies of both non-human mammalian cells and other eukaryotic and prokaryotic species. Study endpoints included chromosomal aberrations, sister chromatid exchange, DNA comet assays, spermchromatin structural assays, micronucleus formation, thymidine kinase assays, assays for DNA oxidative damage, assays for chromosomal deletions, and (in prokaryotes) reversion assays.

Perhaps most relevant to the assessment of genotoxicity to humans are the results of approximately 25 studies looking at a wide variety of human cells either taken from subjects exposed in vivo or from in vitro exposures. The vast majority of these studies indicated that diazinon or its metabolites diazoxon and diethylthiophosphate led to a variety of genetic changes. Very few results were negative. Studies using other animal systems had a larger number of negative findings, but, nevertheless, produced a meaningful number of positive findings for genotoxicity.

A summary of findings from the Tox21 and ToxCast^{TM} is provided below (Table 8.4).

Table 8.4 Summary of Tox21 and ToxCast findings for diazinon and diazoxon

Test	Diazinon ranking	Diazoxon ranking
Metabolic activation	High Cyt 450 activation	High aromatase inbition
Genotoxicity	Low P53 activation	Negative P53 activation
Epigenetic alterations	Negative	Negative
Oxidative stress	Low ranking oxidative stress marker	Negative
Chronic inflammation	Negative	Negative
Receptor-mediated effects	Intermediate aryl hydrocarbon receptor; low estrogen and nuclear receptors	Negative
Cytotoxicity and cell proliferation	Low cytotoxicity	Negative

8.2.3.4 Summary of Findings

The Working Group concluded that evidence for carcinogenicity in human epidemiology and controlled animal studies was limited. However, members noted the strength of evidence of genotoxicity to human and other animal cells, both from examination of human cells taken from individuals exposed to diazinon as well as in vitro studies of human cells together with evidence from an animal in vivo and in vitro exposures to diazinon. They particularly stressed genotoxic evidence from human in vitro and other mammalian cells both in vivo and in vitro. This included data showing that diazinon induces oxidative stress. In sum, diazinon was classified as a Category 2A substance, a probable carcinogen.

8.2.4 Tetrachlorvinphos

Tetrachlorvinphos is not a widely used pesticide either in the US or globally. In the US applications for plants protection ceased in 1987, and tetrachlorvinphos is now allowed only to protect various livestock that produces such commodities as milk and eggs. It is also used for the treatment of animals, including household pets, for insect infestations [16]. Accordingly, little information is available from epidemiology or animal studies that might be employed to specifically assess its potential carcinogenicity.

8.2.4.1 Epidemiology Findings

The IARC Working Group considered one cohort study. While the study did not examine the effects of tetrachlorvinphos alone, it did look at cancer endpoints among a cohort of workers in a cigarette factory who were treating tobacco leaves with several chemicals including tetrachlorvinphos. The one finding of significance was an elevated Standardized Mortality Ratio for non-Hodgkin's lymphoma among women, especially those with an employment history $\geq$ 15 years. Other findings were not significant, and it was not possible to attribute the one specific finding exclusively to tetrachlorvinphos exposure. The study results did not address the confounding effect of smoking [16].

Three case–control studies were evaluated. The first examined the relationship between lymphopoietic cancers among male farmers in Minnesota and Iowa. While this study did get detailed exposure details from farmers and used population-based data, the number of incident cases was small, the levels of exposure were unknown, and the findings were insignificant [16].

The second used pooled data from three case–control studies. While the investigation adjusted for confounders such as age, state, and responder type, it did not adjust for multiple pesticide use. The study did not report findings of significance. Exposure intensity was not known [16].

The third case–control study re-examined the earlier pooled study which focused on farmers in Minnesota, Kansas, and Nebraska. This examination used statistical techniques, including hierarchical regression to adjust for the effects of exposure to pesticides other than tetrachlorvinphos. These findings were also not significant [16].

8.2.4.2 Controlled Exposure Animal Studies

The IARC Working Group considered three controlled exposure studies of mice and three controlled exposure studies of rats. All mice studies involved $B6C3F_1$ mice. The first reported in 1978 by NTP (NTP 1978b) found a significant increase in trend for hepatocarcinoma and adenoma among treated males as well a significant incidence of hepatocarcinoma and adenoma combined among treated males as compared with controls. While the increase in hepatocarcinoma alone was not significant in females, the increase in adenoma and hepatocarcinoma considered together was. No other significant findings were reported.

The second study, which involved a somewhat longer exposure period, also found significant increases in hepatocarcinoma and adenoma in both males and females as well as a significant increase in renal tubule adenoma and carcinoma in males. A later examination of pathology materials by a consultant pathologist confirmed the findings of renal tubule adenoma and carcinoma but classified more of the findings as adenoma rather than carcinoma [16].

A third investigation, reported by EPA, noted significant increases in hepatocarcinoma and adenoma in both males and females [16].

Controlled exposure studies on rats used three strains. The first reported by NTP used Osborne-Mendel rats. The only significant increases were thyroid C cell adenoma and cortical adenoma in females. The second study, reported by EPA, used Porton Wister rats and did not report any findings of significance. The third study, which used Sprague Dawley rats, reported a significant trend in adrenal pheochromocytoma, but no other findings of significance [12].

8.2.4.3 Genotoxic Effects

No studies looking at cells taken from humans exposed in vivo were considered by the Working Group. One in vitro study of cultured human lymphocytes reported chromosomal aberrations. Three investigations of bone marrow taken from Swiss mice exposed in vivo to tetrachlorvinphos looked at micronucleus formation. Two of these reported positive findings. The investigation involving dermal exposure of mice reported negative findings. In vitro findings using non-human mammalian systems involved spleen cells from Swiss mice and Chinese hamster ovary cells. Mice spleen cells exposed to tetrachlorvinphos were found to have chromosomal aberrations and demonstrated sister chromatid exchange. Chinese hamster ovary cells exhibited aberrations if the tetrachlorvinphos had undergone metabolic activation, using a liver activation system.

Table 8.5 Summary of Tox21 and ToxCast findings for Z-tetrachlorvinphos

Test	Ranking
Metabolic activation	No findings in cell-free assays
Genotoxicity	Very moderate activity
Epigenetic alterations	Active for DNA binding although this may represent oxidative stress
Oxidative stress	Intermediate activity
Chronic inflammation	Very low activity
Receptor-mediated effects	Intermediate to high activity
Cytotoxicity and cell proliferation	Low to moderate

No prokaryotic cells studied were found to have genotoxic responses to tetrachlorvinphos. This was also true of a yeast strain. One plant system demonstrated chromosomal aberrations.

Tetrachlorvinphos is unique among the four organophosphate pesticides considered by IARC in that it does not undergo oxidation to become a more toxically active compound. Thus, in the Tox21 ToxCast™ high-throughput assays, only tetrachlorvinphos was considered. Additionally, available data from Tox21 ToxCast™ assays were available only for Z-tetrachlorvinphos, the predominant isomer of commercial tetrachlorvinphos.

Information from the tests is summarized in Table 8.5.

8.2.4.4 Summary of Findings

Based on the findings that no epidemiologic studies found tetrachlorvinphos to be carcinogenic, but some evidence of carcinogenicity in controlled animal exposure studies and some findings of genotoxicity, the Working Group classified tetrachlorvinphos as a Category 2B substance, a possible carcinogen.

8.2.5 Glyphosate

In its evaluation of glyphosate as a carcinogen, the IARC Working Group considered findings from epidemiology, controlled animal exposure studies, and genotoxic data, as they had done for their evaluation of the four organophosphate pesticides. However, information on glyphosate was not available in the high-throughput ToxNET™ program, so this could not be included in the final deliberations.

8.2.5.1 Epidemiology Findings

The one cohort study considered by the Working Group was the AHS. Studies looking at cancer endpoints considered findings for applicators, as well as their spouses and children, and considered a variety of cancers. However, the early AHS studies, while looking at a variety of chemicals applied to crops, did not specifically assess glyphosate use, although a study published in 2004 did consider its effects. Findings from the study were not significant [9]. From the AHS studies as a whole, results were not significant for a variety of other cancers, including breast cancer (generally of spouses of applicators) as well as childhood cancers. Non-Hodgkin's lymphoma was considered in these earlier investigations [16]. For these reasons, the Working Group's assessment of glyphosate must be interpreted with caution. It is important to point out that, lacking information about glyphosate from the inception of the AHS study, lifetime exposure estimates or intensity of lifetime exposures may not have been available for a meaningful period of time.

Fourteen case–control studies looking at the effects of glyphosate were considered by the IARC Working Group. While it was not the only endpoint considered, a large number did look at non-Hodgkin's lymphoma cases as well as closely related lymphopoietic cancers. In the US, the studies focused on cancer cases in farming states, including Iowa, Minnesota, Nebraska, and Kansas. While many findings reported point estimates greater than 1, findings were not significant. An exception to this was an examination of pooled data which used both logistic regression and hierarchical regression to adjust for multiple pesticide exposures. This study reported a significant increase using logistic regression, but a non-significant increase with exposure using hierarchical regression [10]. When asthma was considered as an effect modifier, findings from glyphosate exposure were found to be non-significant.

Three Canadian studies were considered. A cross-Canada study, examining the period 1991–1994 with exposure as a dichotomous variable reported non-significant findings. Another study, looking at Hodgkin's lymphoma, non-Hodgkin's lymphoma, and multiple myeloma in six Canadian provinces, reported non-significant results. This was also true of a later study in the same provinces [16].

Epidemiologists in several European locations also investigated the effects of glyphosate. While two of the three published studies conducted in Sweden did not report significant findings for several hematopoietic cancers, one investigation did report a significant increase in non-Hodgkin's lymphoma when using univariate analysis. Another study, reported in 2008, noted significant findings for non-Hodgkin's lymphoma as well as aggregate lymphomas of different types among those exposed to glyphosate for more than ten years. A French study looking at data from 2000 to 2004 reported non-significant findings. Another study conducted in several European countries reported no significant findings [16].

8.2.5.2 Controlled Exposure Animal Studies

The Working Group considered two controlled exposure mouse studies as providing adequate data. Both involved CD1 mice. The first study, a 24-month study conducted by EPA, found a significant positive trend of renal tubule adenoma in male mice. A re-analysis of the tissue samples found a significant trend with respect to exposure for renal tubule carcinoma and renal tubule adenoma and carcinoma combined [14]. The Joint FAO/WHO Meeting on Pesticide Residues (JMPR) reported a significant increase in hemangiosarcoma in male mice, and a non-significant increase in females [17].

While results in controlled exposure studies in rats did not consistently report significant trends in tumor incidence, some findings were significant. An investigation using Sprague–Dawley rats noted significant differences in the incidence of pancreatic islet cell adenoma in a pairwise comparison with control male rats in the lowest dose and highest dose categories. The same study reported a significant trend in the development of hepatocellular adenoma in males and thyroid follicular adenoma in females. However, there were no findings of progression to carcinoma. EPA reported that male Sprague–Dawley rats had increases in pancreatic islet adenoma. In pairwise comparisons, the development of the adenoma was significant in the lowest dose group, but the trend was non-significant [16].

8.2.5.3 Genotoxicity

Because no data were available for glyphosate in the ToxNET™ program, it was not possible to develop a ToxPi profile. However, a large body of information about the genotoxicity of glyphosate, its metabolite AMPA, as well as glyphosate-based formulations, were available for the Working Group's assessment. The human genotoxicity investigations included four studies reporting genotoxicity in cells taken from humans having in vivo exposures; ten in vitro studies looking at the effects of glyphosate on a variety of cells; two in vitro studies looking at genotoxicity of AMPA; and five in vitro studies of genotoxicity for cells exposed to glyphosate-based formulations. The most commonly studied cells were lymphocytes, and the most common endpoints of genotoxicity were chromosomal aberrations, the DNA comet assay, or micronucleus formation. The vast majority of endpoints were positive [16].

Cells taken from non-human mammals exposed in vivo to glyphosate, AMPA, or glyphosate-based formulations were studied in 22 investigations. Endpoints for genotoxicity included the formation of DNA adducts, chromosomal aberrations, and micronucleus formation. Ten in vivo studies looked at glyphosate. Results were evenly divided between statistically positive and negative findings. The one study looking at AMPA was positive, while cells from animals exposed in vivo to glyphosate formulations yielded five significantly positive responses, five negative, and one equivocal response. Cells from non-human mammals studied in vitro included hepatocytes from Fisher 344 rats, ovary cells from Chinese hamsters, and bovine lymphocytes exposed to glyphosate; Chinese hamster ovary cells exposed to AMPA; and bovine lymphocytes exposed to glyphosate formulations. An equal number of negative and significant endpoints was reported: two for sister chromatid exchange and one for micronucleus were significant [16].

Given its widespread use and resulting presence in the environment, it is not surprising that a large number of in vivo and in vitro investigations looking at non-mammalian species were considered by the IARC Working Group. Numerous fish species were studied. In addition, the response of cells taken from exposed caiman lizards, frogs, tadpoles, snails, oysters, clams, mussels, earthworms, drosophila, and various plant species was assessed. Of the 55 in vivo exposures, 13 involved glyphosate exposure, two AMPA, and the remainder used glyphosate formulations with the comet assay, micronucleus, and chromosomal aberrations as the primary endpoints. Among these assessments, 41 reported significant genotoxicity. Results of in vitro studies that used prokaryotic cells, were, in contrast, equivocal or negative [16].

In addition to the specific tests for genotoxicity, a number of studies looked at oxidative stress. This included in vitro studies of oxidative stress in a number of human cell lines as well as in vivo and in vitro studies of non-human mammals and other animals. Exposures were to glyphosate, AMPA, and glyphosate-based formulations. While not all studies reported positive findings of stress, the vast majority did. Some of these findings were "challenged" by adding antioxidants to reaction mixtures to ascertain whether the oxidative stress effects were diminished.

8.2.5.4 Summary of Findings

The IARC Working Group concluded that limited information about glyphosate carcinogenicity in human subjects was available. They also concluded that controlled exposure animal studies provided sufficient evidence of carcinogenicity. In addition, the large body of information both about the genotoxicity of glyphosate as well as information about glyphosate's capacity to induce oxidative stress (an important step in carcinogenicity) supported the overall conclusion of the Working Group, namely that glyphosate be classified as a Category 2A, a probable carcinogen.

References

1. Alavanja MCR, Sandler D, McMaster SB, Zahm SH, McDonnell CJ, Lynch CF, Pennybacker M, Rothman N, Dosemeci M, Bond AE, Blair A (1996) The Agricultural Health Study. Environ Health Perspect 104(4):362–369. DOI: https://doi.org/10.1289/ehp.96104362
2. Bai Y, Zhou L, Wang J (2006) Organophosphorus pesticide residues in market foods in Shaanxi area, China. Food Chem 98(2):240–242
3. Blair A, Grauman D, Lubin JH, Fraumeni JF (1983) Lung cancer and other causes of death among licensed pesticide applicators. J Natl Cancer Inst 71(1):31–37
4. Burnam W (1984) Ethyl parathion, review of chronic/oncogenic rat study. T. Branch. Washington, DC, US Environmental Protection Agency

5. Cabello G, Valenzuela M, Vilaxa A, Durán V, Rudolph I, Hrepic N, et al (2001) A rat mammary tumor model induced by the organophosphorous pesticides parathion and malathion, possibly through acetylcholinesterase inhibition. Environ Health Perspect 109(5):471–479. DOI: https://doi.org/10.1289/ehp.01109471
6. Coble J, Thomas K, Hines CJ, Hoppin JA, Mustafa D, Curwin B, Lubin JH, Freeman LEB, Blair A, Sandler DP, Alavanja MCR (2011) An updated algorithm for estimation of pesticide exposure intensity in the agricultural health study. Int J Environ Res Public Health 8(12):4608–4622. DOI: https://doi.org/10.3390/ijerph8124608
7. Darko G, Akoto O (2008) Dietary intake of organophosphorus pesticide residues through vegetables from Kumasi, Ghana. Food Chem Toxicol 46(12):3703–3706. DOI: https://doi.org/10.1016/j.fct.2008.09.049
8. Davis JR, Brownson R, Garcia R, Bentz BJ, Turner A (1993) Family pesticide use and childhood brain cancer. Arch Environ Contam Toxicol 24(1):87–92. DOI: https://doi.org/10.1007/BF01061094
9. De Roos AJ, Blair A, Rusiecki JA, Hoppin JA, Svec M, Dosemeci M, et al (2005) Cancer incidence among glyphosate-exposed pesticide applicators in the agricultural health study. Environ Health Perspect 113(1):49–54. DOI: https://doi.org/10.1289/ehp.7340
10. De Roos AJ, Svec M, Blair A, Rusiecki JA, Dosemeci M, Alavanja MC et al (2005) Glyphosate results revisited: De Roos et al. respond. Environ Health Perspect 113(6):A366–367
11. De Roos AJ, Zahm S, Cantor KP, Weisenburger DD, Holmes FF, Burmeister LF, et al (2003) Integrative assessment of multiple pesticides as risk factors for non-Hodgkin's lymphoma among men. Occup Environ Med 60(9). DOI: https://doi.org/10.1136/oem.60.9.e11
12. EPA (1995) Peer review of Mulhern M, et al (1993) Sprague-Dawley rat carcinogenicity study. US EPA Office of Pesticides and Toxic Substances. Washington, DC, US Environmental Protection Agency
13. EPA (1997) Malathion: 2-year chronic feeding/carcinogenicity study in Fischer 344 rats. US EPA Office of pesticides and toxic substances, Washington, DC
14. Eustis SL, Hailey J, Boorman GA, Haseman JK (1994) The utility of multiple-section sampling in the histopathological evaluation of the kidney for carcinogenicity studies. Toxicol Pathol 22(5):457–472. DOI: https://doi.org/10.1177/019262339402200501
15. Guha N, Guyton K, Loomis D, Barupal DK (2016) Prioritizing chemicals for risk assessment using chemoinformatics: examples from the IARC monographs on pesticides. Env Health Perspect 124(12):1823–1829. https://doi.org/10.1289/EHP186
16. IARC (2017) "IARC Monographs on the Evaluation of Carcinogenic Risks to Humans. Some organophosphate insecticides and herbicides/IARC working group on the evaluation of carcinogenic risks to humans" volume 112
17. JMPR (2006) Glyphosate. In: Joint FAO/WHO meeting on pesticide residues, pesticide residues in food. Geneva World Health Organization, pp 95–169. https://www.who.int/foodsafety/jmprsummary2016.pdf
18. McDuffie HH, Pahwa P, McLaughlin JR, Spinelli JJ, Fincham S, Dosman JA, et al (2001) Non-Hodgkin's lymphoma and specific pesticide exposures in men: cross-canada study of pesticides and health. Cancer Epidemiol Biomarks Prev 10(11):1155–1163
19. NIH (2015) Questionnaires and study data. Agricultural Health Study. National Institutes of Health. Washington DC, National Institutes of Health. https://aghealth.nih.gov/collaboration/questionnaires.html
20. NTP (1978) Bioassay of malathion for possible carcinogenicity. Study No. NCI-CG-TR-24. Natl Cancer Inst Carcinog Tech Rep Ser 24:1–102
21. NTP (1979) Bioassay of malathion for possible carcinogenicity (CAS No. 121-75-5). Natl Toxicol Program Tech Rep Ser 192:1–87
22. NTP (1979) Bioassay of parathion for. possible carcinogenicity. Natl Cancer Inst Carcinog Tech Rep Ser 70:1-123.
23. NTP (1979) Bioassay of diazinon for possible carcinogenicity. Natl Cancer Inst Carcinog Tech Rep Ser.137:1-115

24. Rinde E (1989) Second peer review of parathion. Science Analysis and Coordination Branch. United States Environmental Protection Agency, Washington, DC.
25. Settimi L, Costellati L, Naldi M, Bersani G, Olanda S, Maiozzi P (1999) Mortality among workers in an Italian cigarette factory. Occup Med (Lond) 49:361–364. DOI: https://doi.org/10.1093/occmed/49.6.361
26. Slauter, RW. 18 month oral (dietary) oncogenicity study in mice: Malathion. Lab project No. 668-001. Unpublished study prepared by International Research and Development Corporation, Mattawan, MI. EPA MRID 43407201. Toxicological Profile for Malathion; U.S. Department of Health and Human Services, Agency for Toxic Substances and Disease Registry, Public Health Service: Atlanta, 1994
27. Smith MT, Guyton K, Gibbons CF, Fritz JM, Portier CJ, Rusyn I, DeMarini DM, Caldwell JC, Kavlock RJ, Lambert PF, Hecht SS, Bucher JR, Steward BW, Baan RA, Cogliano VJ, Straif K (2016) Key characteristics of carcinogens as a basis for organizing data on mechanisms of carcinogenesis. Environ Health Perspect 124(6):713–721. https://doi.org/10.1289/ehp.1509912
28. Straif K, Grosse Y, Lauby-Secretan B, El Ghissassi F, Bouvard V, Benbrahim-Tallaa L, Guha N, Mattock H (2014). Future priorities for the IARC monographs. The Lancet 15(June):p683–p684. DOI: https://doi.org/10.1016/S1470-2045(14)70168-8
29. Tice RR, Austin C, Kavlock RJ, Bucher JR (2013) Improving the human hazard characterization of chemicals: a tox21 update. Environ Health Perspect 121(7):756–775. DOI: https://doi.org/10.1289/ehp.1205784
30. Zendzian R (1989) Parathion, chronic/oncogenicity study in rats. US EPA Office of Pesticides and Toxic Substances. Washington, DC, United States Environmental Protection Agency

Part III
Implications of and Responses to IARC's Classification of Parathion, Malathion, Diazinon, Tetrachlorvinphos and Glyphosate

Chapter 9
Hazard Identification: Its Relationship to Risk Assessment, Risk Perception, and Regulation

Abstract The purpose of the IARC Monograph series is to identify compounds and substances that pose a carcinogenic risk and to classify the risk according to the risk level. Essentially this is hazard identification. Hazard identification is the first step in the development of a quantitative risk assessment, which may be developed for different scenarios. This chapter will discuss these distinctions and their implications. Generally, risk assessment establishes a quantitative measure of how much risk is involved from a given level of exposure, and, based on this, defines what is considered a safe level or safe range of exposure. While working from hazard identification, risk assessment considers other issues such as conditions of exposure, the duration of exposure, the make-up of the exposed population, as, for example, workplace exposure vs. continuous exposure, exposure to the general public as opposed to the workplace, and other factors such as the needs of vulnerable groups, or years of quality life that may be lost. As is true with hazard identification, risk assessments can be modified as new information becomes available. These processes and issues will be specifically discussed with respect to the compounds identified as probable or potential carcinogens by IARC and published in Monograph 112. The discussion will also address such questions as risk assessment for the general public vs. risk assessment for those who are occupationally exposed.

9.1 Regulatory Frameworks Affecting Use and Ban of Pesticide and Herbicide Use

The pesticides and herbicides reviewed by the IARC Working Group in 2015 were, when introduced, widely used. However, in numerous locations, as more information has became available, the use of many has either been banned or highly restricted. Often this has been true globally. Nevertheless, the terms "ban" and "restricted use" are not necessarily subject to the same regulations in various countries or geographic locales. Because of these differences as well as differences in enforcement mechanisms, it is probable that certain substances will continue to be used in many locales.

M. Richmond, *Cancer Hazards: Parathion, Malathion, Diazinon, Tetrachlorvinphos and Glyphosate*, AESS Interdisciplinary Environmental Studies and Sciences Series,
https://doi.org/10.1007/978-3-030-81953-8_9

An examination of aspects of the regulatory framework provides a useful lens for considering aspects of how this may "play out" in the global arena.

Globally, the Rotterdam Convention, a multinational treaty under the auspices of the United Nations, encourages multinational participation in the regulation of hazardous chemicals, including pesticides and herbicides. Among the roles that signatory nations can play in regulating the import of chemicals (allow or ban), member parties can also participate in cooperative actions by informing the Secretariat of the Rotterdam Convention of a decision to ban or restrict a chemical, based either on human health or environmental concerns. These decisions are published on a regular basis. Countries exporting chemicals are required to certify that organizations producing these chemicals within their geographic boundaries do not export such chemicals to countries banning import. In 2018 the Rotterdam Convention had 161 parties, including 158 UN member nations as well as the EU, Cook Islands, and Palestine. The US, a UN member nation, does not participate in the Rotterdam Convention. Ironically, the US is, however, one of the major food producers as well as pesticide users on a global basis.

A comprehensive article by Donley [5] compared the regulatory framework of four countries or locales that produce the bulk of the world's food supply, and are also collectively the world's largest pesticide users, namely the US, the European Union (EU), Brazil, and China. Interestingly, the article noted that "The USA, EU, China, and Brazil each have separate and distinct pesticide regulatory systems designed to protect, to varying degrees, humans and the environment" but pointed out the significant ways that these regulatory frameworks vary, and the potential impact that this variability has on both the ban and use of specific pesticides [5]. While the article does not specifically address the pesticides considered in the IARC 2017 monograph, the principles outlined in the article, nevertheless, apply to their regulation.

In the US, the EPA is the major regulatory agency that considers each of these substances as they may affect the general public. Depending on the major toxic effects of a given substance, regulation may be based on data regarding acute toxicity or chronic toxicity. The Federal Insecticide, Fungicide, and Rodenticide Act (FIFRA) defines the specific role that EPA plays in governing licenses for the use of insecticides, which focuses primarily on the environment. In this context, FIFRA defines the term "unreasonable adverse effects on the environment" to mean: "(1) any unreasonable risk to man or the environment, taking into account the economic, social, and environmental costs and benefits of the use of any pesticide, or (2) a human dietary risk from residues that result from a use of a pesticide in or on any food inconsistent with the standard under Sect. 408 of the Federal Food, Drug, and Cosmetic Act" [7].

In general, rather than banning the use of a given pesticide, the specific EPA licensing requirements for its use, especially for a pesticide whose use is decreasing because of generic substitutions, may make it impractical for a commercial producer to continue producing that substance. In effect, this means that the pesticide re-registration application is withdrawn. However, voluntary withdrawal is a gradual process and it may take up to four years until the pesticide use has effectively ceased. In contrast, if EPA decides to ban a pesticide, the decision must pass through several steps of approval prior to the cancellation. This includes approval from the US Department of Agriculture as well as the FIFRA Scientific Advisory Panel. FIFRA requires that a cost–benefit analysis be conducted for any harm not associated with exposure to humans through food intake (e.g., harm to water quality, harm to biodiversity, or environmental harm). The industry or company that has registered the pesticide can also request a judicial hearing, and may, in addition, request that the decision be appealed to EPA with the possibility of restricting pesticide use rather than canceling use [5]. Not surprisingly, in some cases, an agreement is made to restrict the use rather than to ban completely. While this may be effective, in essence, it does not remove a particular substance from the market and does not guarantee that it will not be used for other purposes.

In the EU, registration and use requirements may be more stringent in some respects. For the use of a pesticide to be approved, the industry registering the pesticide must demonstrate that the pesticide "can be used in a way that does not result in harm to humans or the surrounding environment" [5]. No pesticide that is identified as a carcinogen, mutagen, reproductive toxin, or endocrine disruptor will be approved or can continue to be used unless human exposure is negligible. However, in contrast to requirements for approval, once a pesticide is approved for use, its removal, unless clearly defined as a carcinogen, mutagen, reproductive toxin, or endocrine disruptor, is less clearly proscribed, although some details are similar to those in the US. As is true in the US, a pesticide can either be withdrawn by the producer or considered withdrawn if it is not properly registered. In the EU registration includes payment of fees as well as submission of necessary safety data. A pesticide can also be banned if the EU government agency deems that it would be appropriately unsafe, although specific details may not be clear.

Brazil and China, the other two nations viewed as major global food producers by Donley [5], have a regulatory approach that could be considered more direct. In theory, a pesticide, once banned, remains banned for use for an unspecified period of time. The primary criterion for the ban is safety, although this is not necessarily clearly defined. It is also true that government structures in these two countries may have a more direct impact on both the processes of approval and banning, and that such decisions have the potential to be adopted or revoked with fewer structures involved in these processes.

9.1.1 Overview of Regulatory Issues for Substances Considered by IARC in 2015 as Probable Carcinogens

Shortly after IARC's announcement of the agency's classifications of malathion, diazinon, and glyphosate as probable carcinogens (but prior to the publication of Monograph 112), an expert task force was created to examine whether the use of each of these substances should be modified in light of the classifications [11].

The initial recommendation [11], published in September 2015, stated that:

> To date, WHO has not made any changes to the technical instructions for the use of diazinon and malathion in public health, nor for glyphosate. Accordingly, the current technical recommendations remain in effect.

The final task force, a joint FAO/WHO Meeting on Pesticide Residues (JMPR), met in 2016. Of particular concern at the time was the specific need to control the vector for the Zika virus as well as the role that malathion spraying has had to control mosquito vectors for malaria and dengue. These concerns were examined in the context of whether vector control use would place consumers at risk from exposure to malathion residues in food. Following deliberations, the task force developed guidelines.

According to Pesticide Action Network International (PAN International), in 2019 the majority of countries in the EU had restrictions of some type for the use of diazinon, and a total of 32 countries across a spectrum of locales had the same types of action. In contrast, the use of malathion was far less restricted, with only two countries globally having some type of restriction. In 2019 PAN International did not report any restrictions on the use of glyphosate [12]. However, this has recently changed, as noted below.

9.1.1.1 Details of Malathion Regulation

The guidelines for malathion use in vector control specifically address a malathion formulation that would have been used in spraying to control vectors, Malathion EW440 G/L [11, 16, 17]. These are summarized as follows:

> Assuming that: malathion EW440 G/L formulation is used for outdoor space spraying only, and that it (1) complies with the WHO specification for impurities, (2) strictly follows guidelines for space spraying to control vectors (3) hands and arms of the applicator are covered with suitable protection (4) application is no more than 352 g/ha, the WHO risk assessment for indoor and outdoor space spraying of insecticides predicts that the systemic dose of malathion to operators will likely exceed acceptable levels if not strictly adhered to.
>
> However, if applicators comply, the acceptable dose will be less than the long-term dose by 63%, and 22% of the toxic dose for acute exposure for applicators with hand-held devices, and 29% of the long-term dose and 11% of the toxic dose for acute exposure for vehicle-mounted spraying operation.

Other conditions specify the need to address bystander exposures, both in terms of long-term and acute exposure as well as prevention of untoward effects on the aquatic environment, soil organisms, birds, and herbivores [16, 17].

9.1.1.2 Details of Diazinon Regulation

As noted earlier, the joint FAO/WHO group did not direct any changes in the use of diazinon. A JMPR publication issued in 2016 noted that diazinon, in amounts that would be taken in through food, was not of concern, and did not feel that regulations should be changed [16, 17]. The US has banned or not registered diazinon for personal household and gardening use. This is true for a number of other countries. This change has significantly lowered the amount of diazinon in the environment, which indirectly may affect dietary intake and other types of exposure. Wang and coworkers, in a 2017 publication [15], note the meaningful decrease of diazinon in California surface waters after its residential use was canceled.

9.1.1.3 Details of Glyphosate Regulation

As cited earlier, the response regarding glyphosate has been strong. To some extent, this reflects the views of specific organizations. Some of the response has been heavily politicized and polarized.

A joint publication by the Pan American Health Organization and WHO in September 2015 (following findings by IARC in 2015) noted that, at that time, the joint finding of WHO and the Food and Agriculture Organization (FAO) felt no need to alter its recommendations regarding the use of glyphosate that might have an impact on public health [11]. This decision was based on food intake rather than other types of exposure.

Nevertheless, as of July 2020, a number of countries have either banned or restricted the use of glyphosate [2]. The listing is provided in Table 9.1. At the same time, it is notable that many others have not. For example, with the exception of Malawi, no Saharan or sub-Saharan African country has banned glyphosate, nor is glyphosate banned in China, Korea, or Japan, nor is its use curtailed or banned in Russia.

While neither curtailed nor banned in the US as a whole, communities or geographic areas in 25 states have enacted ordinances, resolutions, or legislation banning or curtailing glyphosate's use. In Vermont, several bills banning use have been introduced to the legislature [2]. In New York, a bill banning the use of glyphosate-containing herbicides on all state lands was passed in July 2020 and is awaiting the governor's signature [3].

At the same time, both in the US and internationally, a number of government agencies have taken different approaches to glyphosate regulation. In part, these approaches are based on the differences between hazard identification and various risk assessment models [6]. Many agencies have developed risk assessment models that focus on risk for the general population. Such models consider the major source of exposure to be the dietary intake of residues or metabolite residues in food sources exposed to glyphosate. An additional consideration has been agency access to information that can be used for risk assessment by different regulatory agencies. EPA, WHO/FAO, and the European Food Safety Authority (EFSA) have had access to and incorporated data from industry studies assessing glyphosate. These data were not available in the peer-reviewed literature evaluated by the IARC Working Group.

At the same time, however, some of the differences derive from different evaluations of studies used by IARC in its 2015 decisions. An early illustration was the disparities between IARC's classification and EFSA's classification. In 2015, based on a report provided by the German Federal Institute for Risk Assessment (BfR),

Table 9.1 Countries where glyphosate is banned or restricted

Country	Type of restriction	Effective date	Other details
Argentina	>30,000 health care professionals and 400 towns or municipalities restrict the use	2015 (health care professionals)	On-going process
Australia	Municipalities and school districts investigating alternatives to glyphosate use	No specific date	On-going process
Austria	Originally banned to take effect in 2020. Citing a technicality, "caretaker" leader has not signed the ban	2019	On-going
Bahrain	Banned use	No date given	Ban also adopted by five other countries in the Gulf Cooperation Council
Belgium	Banned individual use of glyphosate; glyphosate use is completely banned in Brussels	2017	
Bermuda	Banned personal and commercial sale of glyphosate-containing herbicides. Restriction modified to permit use in clearing weeds from the roadside	2017	
Brazil	Originally banned the registration of new products containing glyphosate	2018	Based on findings by Brazilian health agency, ban overturned in 2019
Canada	8 of 10 provinces restrict use. Vancouver banned all use other than use for invasive weeds. New Brunswick is investigating regulations	2019	Not considered a carcinogen by Canadian Pest Management Regulatory Agency

(continued)

Table 9.1 (continued)

Country	Type of restriction	Effective date	Other details
Columbia	Outlawed use of glyphosate to destroy coca (plant producing cocaine)	2015	President asked for lifting of aerial spraying ban, 2019. Judiciary has denied this request noting the burden of proof regarding safety is that of president
Costa Rica	Use prohibited in 11 protected wild areas	2019	
Czech Republic	Banned glyphosate as a weed killer and a drying agent	2019	
Denmark	Government recognizes as a carcinogen. Private use ban in Aalborg	2017	Use banned on all post-emergent crops to prevent contamination in food, 2018
El Salvador	Banned because of links to kidney disease	2013	Appeared to "stall' in 2016
Fiji	All use banned	2021	
France	With limited exceptions, Roundup™ use will be eliminated	2019	With few exceptions, all glyphosate use will be eliminated in 2021
Germany	Glyphosate use will be banned	2023	Presently not considered a carcinogen by German Federal institute for occupational safety and health
Greece	Voted against re-licensing glyphosate	2017	March 2018 government approved a 3-year license renewal for Roundup™
India	Punjab banned sale of any glyphosate formulations Kerala banned use, sale, and distribution of glyphosate	February 2018 October 2019	
Italy	Banned as harvest pretreatment and placed other restrictions on use	2016	In 2017, one of 7 EU nations to vote against re-licensing use in EU

(continued)

Table 9.1 (continued)

Country	Type of restriction	Effective date	Other details
Kuwait	Banned	No date given	Ban also adopted by five other countries in the Gulf Cooperation Council
Luxembourg	Banned	Market authorization withdrawn February 1, 2020 Remaining stocks can be used until June 30, 2020 All use must stop by December 31, 2020	First EU nation to issue a total ban
Malawi	Licenses for glyphosate-containing products, e.g., Roundup™ immediately suspended	April 2019	
Malta	Glyphosate spraying in all public places banned	July 2019	
Mexico	Gradually phasing out glyphosate use with ban in 2024	2024	
Netherlands	Ban on all commercial use of glyphosate	No date given	
New Zealand	Physicians and scientists for global responsibility called for total ban	2015	Auckland and Christchurch advocate for reduction of chemicals in weed control Not considered a carcinogen by New Zealand Environmental Protection Agency
Oman	Banned	Announced 2016	Member of Gulf Cooperation Council
Portugal	Prohibits the use of glyphosate in all public places	No date given	President of Portuguese Medical Association calls for worldwide ban on glyphosate use
Qatar	Banned		Member of Gulf Cooperation Council
St. Vincent and the Grenadines	Immediate suspension of all glyphosate imports	No date given	

(continued)

Table 9.1 (continued)

Country	Type of restriction	Effective date	Other details
Saudi Arabia	Banned		Member of Gulf Cooperation Council
Scotland	Five of six members of EU parliament voted to phase out glyphosate by 2022	2017	Edinburgh and Aberdeen phasing out use
Slovenia	Letter to EU advocating for an exit plan for glyphosate use	2018	
Spain	Banned in Barcelona, Madrid, Zaragoza, and the region of Extremuda	No date	
Sri Lanka	First country to enact a total ban	No date	Ban lifted in 2018 because of overgrowing weeds and crop losses
Sweden	Users permitted to use only "low risk substances" (not including glyphosate)	2017	Issued by Swedish Chemicals Agency
Switzerland	Green Party proposed a plan to ban glyphosate	2017	Plan rejected by Swiss Federal Council Two supermarkets have removed all glyphosate-containing products from shelves
Thailand	Deputy Agricultural Minister stopped licensing glyphosate as a farm chemical	August, 2019	Reduced to a restriction, December 2019 following industry pressure
United Arab Emirates	Banned	No date	Member of Gulf Cooperative Council
United Kingdom	Homebase announced it would review sale of Roundup™	August 10, 2018	Examination of sales indicates that sales continue
Vietnam	Announcement of removal of glyphosate as a pesticide that could be imported that would take place "in the near future."	2019	Action unclear

EFSA concluded that "glyphosate is unlikely to pose a carcinogenic hazard to humans and the evidence does not support classification with regard to its carcinogenic potential" [9]. In reaching this conclusion, EFSA commented on IARC's assessment of human epidemiology studies, questioning the conclusions reached by the IARC working group regarding several case–control studies. EFSA also questioned the overall assessment of several animal studies and noted that genotoxicity studies seen by IARC were "overwhelmingly negative" [9]. In response to the EFSA decision, Portier and 94 other authors published a response [13]. The response, which contains specific answers to each of the points made by the EFSA renewal assessment report provided by BfR, also notes that significant amounts of information made available to BfR were not open, transparent, and available to IARC working group members. In its general concluding comments, the response notes that:

> Science moves forward on careful evaluations of data and a rigorous review of findings, interpretations and conclusions. An important aspect of this process is transparency and the ability to question or debate the findings of others [13].

Replying to the commentary published by Portier et al., Tarazona and co-authors [14] note that the toxicology reference value (the biological indicator value for a substance for a defined group) for the general population is quite low, and, based on this information, glyphosate is not a "public concern." The debate regarding EFSA is continuing.

In the US, EPA published an interim decision in January 2020 [8] noting that:

> After a thorough review of the best available science, as required under the Federal Insecticide, Fungicide, and Rodenticide Act, EPA has concluded that there are no risks of concern to human health when glyphosate is used according to the label and that it is not a carcinogen.

EPA was expected to issue the final report in the fall of 2020. It has not yet been released.

In August 2020, the ATSDR issued a complete toxicological profile for glyphosate which includes toxicity data on cancer and also a number of other health endpoints. At the same time, the report provides extensive details on epidemiologic studies, indicating that for some lympho-hematopoietic cancers, that while the risk was low, exposure did increase cancer risk [1]. The report does make the distinction between exposure to glyphosate alone (which is possible in laboratory animal studies) and exposure to glyphosate and adjuvants, noting that epidemiologic information on human exposure cannot distinguish between glyphosate and glyphosate plus adjuvant effects.

The National Toxicology Program (NTP) has been conducting the on-going toxicology studies of glyphosate examining genotoxicity and oxidative damage both from glyphosate alone and for glyphosate in various formulations. While some preliminary findings have been reported in poster presentations, final results are still pending [10].

In February 2018, most likely in response to letters sent in 2017 to the head of IARC and the Acting Secretary of Health and Human Services by senior members of the Congressional Committee on Science, Space and Technology, the Congressional Committee conducted a hearing. The letters had raised the issue of possible conflicts of interest and transparency in IARC discussions regarding glyphosate. A full transcript of the Committee hearing, available from the US Government Publishing Office or online at http://science.house.gov, details deliberations and testimony by representatives of EPA, Pastoor Science Communications, the National Resources Defense Council (NRDC), and Dr. Robert Tarrone. The former two testimonies supported the assessment that glyphosate does not present a carcinogenic hazard, while the testimony from the NRDC, in support of IARC, likened the tactics used against glyphosate findings to those used by the tobacco industry in questioning tobacco's carcinogenicity. The testimony by Dr. Tarrone raised questions that he and others had posed about IARC analyses of study findings. The chair of the Committee, Rep. Lamar Smith of Texas, in opening remarks, raised questions of "selective use of data and lack of public disclosure" as well as issues of food security to support a proposal to withhold US financial support to IARC. The representative of the minority party on the Committee, raised issues of industrial pressure to counter IARC's classification of glyphosate while affirming the importance of technical innovation in the food sector [4].

While no specific action was taken with regard to funding of IARC, during the Trump administration, the executive branch of the US stated its intention to withdraw from WHO, which, presumably, would have meant the withdrawal of financial or other support from IARC as well. The Biden administration through an executive order has rejoined WHO.

References

1. ATSDR (2020) Toxicological profile for glyphosate. US Department of Health and Human Services, Agency for Toxic Substances and Disease Registry, Atlanta GA
2. Baum Hedlund Law (2020) "Where is glyphosate banned?" https://www.baumhedlundlaw.com/toxic-tort-law/monsanto-roundup-lawsuit/where-is-glyphosate-banned-/?
3. Beyond Pesticides (2020) "New York state legislature bans glyphosate (Roundup) in Public Parks, Bill Goes to Governor for Signature." Beyond Pesticides, Washington, DC
4. Committee on Science, and Technology, House of Representatives, one hundred and fifteenth Congress, Second Session, February 6, 2018. SN 115-46. "In defense of scientific integrity: examining the IARC monograph programme and glyphosate review", House of Representatives. Washington DC, US Government Printing Office. https://docs.house.gov/meetings/SY/SY00/20180206/106828/HHRG-115-SY00-20180206-SD004.pdf
5. Donley N (2019) The USA lags behind other agricultural nations in banning harmful pesticides. Environ Health 18(44). https://ehjournal.biomedcentral.com/articles/10.1186/s12940-019-0488-0
6. Eastmond DA (2016) "Glyphosate hazard and risk assessment: a comparison of the approaches of two international agencies." http://cbns.ucr.edu/faculty/EastmondToxForumglyphosatepresentation2.pdf
7. EPA (2021) Summary of the federal insecticide, fungicide, and rodenticide act. https://www.epa.gov/laws-regulations/summary-federal-insecticide-fungicide-and-rodenticide-act
8. EPA (2020) EPA finalizes glyphosate mitigation. https://www.epa.gov/pesticides/epa-finalizes-glyphosate-mitigation

9. European Food Safety Authority (2015) Conclusion on the peer review of the pesticide risk assessment of the active substance glyphosate. EFSA J 13:4302–4407. DOI: https://doi.org/10.2903/j.efsa.2015.4302
10. National Toxicology Program (2020) "Glyphosate & glyphosate formulations." National Institute of Environmental Health Sciences. https://ntp.niehs.nih.gov/whatwestudy/topics/glyphosate/index.html
11. Pan American Health Organization/WHO (2015) "Questions and answers on the use diazinon, malathion and glyphosate." Pan American Health Organization/WHO. https://www.paho.org/en/topics/vectors-integrated-management-and-public-health-entomology/questions-and-answers-use
12. PAN International Consolidated (2019) "Pan international consolidated list of banned pesticides." Pesticide Action International. https://pan-international.org/pan-international-consolidated-list-of-banned-pesticides/
13. Portier CJ, Armstrong B, Baguley BC, et al (2016) Differences in the carcinogenic evaluation of glyphosate between the International Agency for Research on Cancer (IARC) and the European Food Safety Authority (EFSA). J Epidemiol Community Health 70(8):741–745. DOI: https://doi.org/10.1136/jech-2015-207005
14. Tarazona JV, Tiramani M, Reich H, Pfeil R, Istace F, Crivellente F (2017) Glyphosate toxicity and carcinogenicity: a review of the scientific basis of the European Union assessment and its differences with IARC. Arch Toxicol 91(8):2723–2743. DOI: https://doi.org/10.1007/s00204-017-1962-5
15. Wang D, Singhasemanon N, Goh KS (2017) A review of diazinon use, contamination in surface waters, and regulatory actions in California across water years 1992–2014. Environ Monit Assess 189(310). DOI: https://doi.org/10.1007/s10661-017-6026-z
16. WHO (2016) "Pesticide residues in food 2016." Special session of joint FAO/WHO meeting on pesticide residues. Food and Agriculture organization of the United Nations/WHO. https://www.who.int/foodsafety/areas_work/chemical-risks/JMPR_2016_Report_May.pdf
17. WHO (2016) "Use of malathion for vector control; report of a WHO meeting Geneva, 16-17 May 2016." WHO pesticide evaluation scheme, department of control of neglected tropical diseases, global malarial programme, department of public health, environmental and social determinants of health. https://www.who.int/publications/i/item/9789241510578

Chapter 10
Response to the IARC 2017 Monograph from the Scientific Community, the Lay Press, the Legal Sector, and the Commercial Sector

Abstract Given the commercial importance of Roundup Ready™ genetically modified seed, it is, perhaps, not surprising that the strongest response to IARC Monograph 112 has been to the designation of glyphosate as a Category 2A carcinogen. Response to the IARC classifications of the four organophosphate pesticides has not been as strong. In part, this may be because a number of other pesticides can serve the same purposes. Parathion has been withdrawn from use in a large number of countries, so the IARC designation has little economic importance. Many countries have also taken significant steps to stringently regulate or curtail the uses of the other organic phosphates. This is especially true in the EU. Certain targeted uses, nevertheless, do remain, especially for malathion, which has been widely used both agriculturally and in insect vector control. Despite these specific considerations, a number of controversies, both about IARC's approach to developing the monographs and implications of the classifications remain. This includes debates within the scientific community, the commercial sector, and the lay press.

10.1 Debates Within the Scientific Community

10.1.1 General Issues

While the bulk of response to the IARC 2017 Monograph classifications has focused on glyphosate's classification, other criticisms regarding IARC's approach have also been raised. Some of these, presented in more general terms, were discussed earlier in Chap. 7. While not specifically directed to one of the five compounds discussed in Monograph 112, these seemingly have grown out of concerns in the scientific community about the IARC risk identification process, which, by some, is regarded as ineffective. A commentary prepared by Boobis et al. [4] maintains that "the IARC classification [is] based on a concept developed in the 1970s that divided chemicals into two classes: carcinogens and non-carcinogens [and that]....this...places into the same category chemicals and agents with widely differing potencies and modes of action." To replace this, the authors advocate an approach based on the nature of the

M. Richmond, *Cancer Hazards: Parathion, Malathion, Diazinon, Tetrachlorvinphos and Glyphosate*, AESS Interdisciplinary Environmental Studies and Sciences Series,
https://doi.org/10.1007/978-3-030-81953-8_10

substance in question: whether it is an acute toxin for which hazard identification can play an important role or a substance that, under certain circumstances, may be toxic, but is not an acute toxin. For the latter group of substances, a risk assessment can play a more prominent role [4].

10.1.2 Organophosphates

10.1.2.1 Response from the Scientific Community

The use of high-throughput ToxCast/Tox21 programs to supplement evaluations from other mechanistic data for the four organophosphate pesticides, described in Chap. 8, was critically discussed in two recent publications. The first, [3], used multiple statistical analyses and machine learning algorithms to evaluate the ToxCast/Tox21 data, both not correcting and correcting for the cytotoxicity burst effect [9]. This is an effect attributed to an adverse cell response when a cell is exposed to a chemical at concentrations close to what would be expected to cause cell death. For test purposes, the ToxCast/Tox21 evaluations used carcinogenic data about substances designated either as positive or negative by EPA. The authors present statistical data suggesting that once corrections for cytotoxic burst are introduced, little or no differences can be found between known carcinogenic substances and those not designated as carcinogens through other assessments using ToxCast/Tox21. It is noteworthy that, while a disclaimer is part of the manuscript, funding was provided by the pro-industry American Chemistry Council, and that the first author, RA Becker, is affiliated with this organization.

Becker and co-authors note that, as their submission was undergoing peer review, Chiu and co-authors [5] submitted a manuscript that discusses the ToxCast/Tox21 approach used to evaluate mechanistic data in several IARC monographs. Specifically, they compare the use of the ToxCast/Tox21 assessment of the four organophosphate pesticides to data from top-ranked carcinogens from other IARC evaluations, using the ten characteristics designated by Smith and co-authors [26] as mechanisms of carcinogenicity relating to hazard identification. Information taken from all of the ToxCast/Tox21 data was presented both graphically and in terms of their ToxPi representation, the visual depiction of the multiple and different in vitro assays resulting in a pie "slice." To compare the ToxPi data for the four pesticides, Chiu and co-authors provide ToxPi data for the top compounds in the ToxCast/Tox21 from other IARC evaluations. The authors found that, with the exception of diazinon, the organophosphate pesticides ranked in the top quartile ToxPi representations. In one category, that of nuclear receptors, all four pesticides ranked in the top quartile.

In concluding remarks, Chiu and co-authors provide the following caveat:

> Until there is better assay coverage, and a standardized methodology for integrating the traditional mechanistic evidence with the data from high-throughput screening platform, it will be difficult to evaluate how predictive high throughput assay data alone can inform conclusions about potential human carcinogenicity. Likewise, different sets of the 10 key

characteristics may be of varying degree of relevance depending on the target organ, as well as on the timing of their alteration along the process leading to a very complex adverse outcome such as cancer.

10.1.3 Glyphosate

The reaction to glyphosate's classification, in contrast to the responses to the organophosphate classifications, has been strong and highly polarized. Many authorities have expressed concerns about the division between public health officials and corporate interests, a concern that was clearly laid out in a commentary by Infante [12] that discusses both IARC's approach to evaluations of substances and agents, and also, in the case of glyphosate, various efforts by industry and pro-industry consultants and lobbyists to discredit IARC. Some controversies have not only appeared in the peer-reviewed literature but were also noted by Infante. The controversy has also developed both in litigation and coverage by the lay press. It continues.

10.1.3.1 Response from the Scientific Community

Among members of the scientific community, response to IARC's classification of glyphosate was strong and, as noted, polarized. Scientists from the agrochemical industry attacked the assessment and, by extension, questioned US government funding for IARC, while scientists affiliated with IARC and a large number of public health experts pointed out IARC's mission, the strength of the working group observations, and the group's conclusions. Others noted the differences between IARC's hazard identification approach and analyses by other expert panels that incorporate hazard identification into the development of risk assessment models that can be used in regulation.

As cited by Richmond [23] criticism from some of those in the scientific community focused on specific aspects of the IARC process that were used to develop the monograph, implying that these were biased. Although cited in an earlier publication [23], to put the following discussion into context, aspects are repeated in this chapter. This includes the following points:

- IARC edited parts of the glyphosate monograph to achieve a specific outcome
- Data from the Agricultural Health Study were deliberately excluded or omitted
- The monograph development was not open to scrutiny and was not transparent
- Working Group members had conflicts of interest and included an invited specialist who was not a member of the Working Group.[1]

[1] As noted in [23], Dr. Christopher Portier, now a consulting scientist, was formerly director or associate director of several US environmental agencies and, while not participating as a member of the expert panel in the IARC evaluation of glyphosate, did attend the meeting. In his present consulting role, he has been an expert witness for a US law firm involved in glyphosate litigation, although at the time he attended the IARC meeting, he was not involved in glyphosate litigation,

In January 2018, IARC issued a detailed response addressing several of these specific points [8]. In the response introduction the agency noted the following:

> Since the evaluation of glyphosate by the IARC Monographs Program in March 2015, the Agency has been subject to unprecedented, coordinated efforts to undermine the evaluation, the program and the organization. These efforts have deliberately and *repeatedly misrepresented the Agency's work. The attacks have largely originated from* the agro-chemical industry and associated media outlets. They have taken place in the context of major financial interests relating to: a) the relicensing of glyphosate by the European Commission; b) hundreds of litigation cases in the USA brought by cancer patients against Monsanto, claiming that their malignancies were caused by glyphosate use; c) and the decision by the Californian Environmental Protection Agency to label glyphosate as a carcinogen (California Office of Environmental Health Hazard Assessment 2017).

The response also clarified the points enumerated above, noting that:

- No parts of the glyphosate monograph were edited to achieve a particular outcome.
- There was no deliberate exclusion of data from the Agriculture Health Study.
- All monograph evaluations are transparent and can be scrutinized.
- IARC has a strong rationale for the inclusion of only publicly available studies in monograph evaluations.
- All Monograph Working Group members who evaluated glyphosate were free from conflict of interests.
- IARC evaluates only agents that have some evidence of carcinogenicity, and the breakdown of evaluated agents is such that roughly half are found not to present evidence of carcinogenicity; 12% have been classified as human carcinogens; and the remaining have been classified as Category 2A (probable) or Category 2B (possible) carcinogens.
- The monographs program re-evaluates an agent when a substantial additional body of scientific evidence becomes available.
- The monograph evaluations place agents in groups according to the strength of evidence of carcinogenicity, not potency.
- IARC monographs identify carcinogenic hazards and do not include a risk assessment.
- IARC evaluations make use of the latest scientific data and methodologies.
- The monographs do not exclude research conducted by industry per se. If findings from industry-conducted studies are published in scientific journals they are considered, if available in sufficient detail to allow independent scientific review. Under the same conditions, the monographs also take account of industry-conducted research in summary form or if placed in the public domain by national regulatory agencies.

---

according to a letter from Reps. Lamar Smith (R-Tex), Andy Biggs (R-AZ), and Frank Lucas (R-OK) to Dr. Christopher Wild, IARC Director, Dr. Portier became involved in glyphosate litigation 9 days after the IARC assessment was announced. A publication by Corporate Europe Observatory [7] "Setting the record straight on false accusations: Dr C. Portier's work on glyphosate and IARC") 2017 defends Dr. Portier's work, noting that he did not sign a contract until 29 days following the IARC meeting, and that more than 90% of his work as an expert witness was "performed and billed" in 2017.

IARC also noted that monograph appraisals take into account "real-world" exposures by evaluating epidemiologic studies. These studies are a central part of evaluations since they deal with people exposed in daily life, including time at work. In addition, when considering scientific evidence of carcinogenicity that includes evidence of biological mechanisms, the working groups place special emphasis on whether the observations are relevant to humans.

As noted in the Infante commentary [12], publications and other activities by industry itself or pro-industry consultants are generally intended to discredit IARC's evaluations. While these were discussed in some detail in an earlier publication [23], because of more recent events, it is informative to include this preceding material to illustrate industrial involvement, in this case, Monsanto. The chronology is as follows.

Before IARC's 2015 hazard assessment, a series of review articles, in part commissioned by Monsanto, was published in the peer-reviewed literature. The articles were written both by individuals associated with Monsanto, other chemical industry representatives, and members of a consortium of approximately 20 industrial organizations working to renew glyphosate registration in the EU. Academicians and private consultants also participated in the series. The focus of each article was a critique of a specific issue expected to be considered by IARC. Included was an evaluation of published animal carcinogenicity studies [11], a critique of data used to assess human genotoxicity and genotoxic potential [17], and details about and evaluations of several unpublished studies that assessed developmental cardiovascular toxicity in animals exposed to glyphosate [18]. Taken together, the conclusion of the reviews was that glyphosate does not present significant genotoxic risks to human populations, and that animal studies do not support a finding of carcinogenic potentials in humans.

Closely following findings of the 2015 working group, a series of five review articles was published in a supplemental edition of Critical Reviews in Toxicology. The foreword to the review articles noted that:

> the Monsanto Company engaged Intertek, a scientific and regulatory consulting firm, to convene an independent scientific panel to evaluate and synthesize the scientific evidence of the potential carcinogenic hazard of glyphosate. The activities and conclusions of the independent panel are reported in the five papers in this special issue. Each of the five papers was rigorously reviewed by 5-10 independent reviewers selected by the CRT [Critical Reviews in Toxicology] Editor and anonymous to the authors. A total of 27 different reviewers participated with several of the individuals reviewing all five papers. The authors of each paper were provided the review comments on their paper and asked to make appropriate revisions. The final papers, published here, represented the work product of the authors. Each paper includes an Acknowledgements section and an extensive Declaration of Interest section [20].

The Critical Reviews in Toxicology supplemental edition included an introductory article [29] that summarized the findings of those participating in the commissioned articles with the following statement:

> The International Agency for Research on Cancer (IARC) published a monograph in 2015 [*Monograph 112 was published in 2017, but findings announced earlier*] concluding that

> glyphosate is "probably carcinogenic to humans" (Group 2A) based on limited evidence in humans and sufficient evidence in experimental animals. It was also concluded that there was strong evidence of genotoxicity and oxidative stress. Four expert panels have been convened for the purpose of conducting a detailed critique of the evidence in light of IARC's assessment and to review all relevant information pertaining to glyphosate exposure, animal carcinogenicity, genotoxicity, and epidemiologic studies. Two of the panels (animal bioassay and genetic toxicology) also provided a critique of the IARC position with respect to conclusions made in these areas. The incidences of neoplasms in the animal bioassays were found not to be associated with glyphosate exposure on the basis that they lacked statistical strength, were inconsistent across studies, lacked dose-response relationships, were not associated with preneoplasia, and/or were not plausible from a mechanistic perspective. The overall weight of evidence from the genetic toxicology data supports a conclusion that glyphosate (including glyphosate-based formulations and aminomethylphosphonic acid) does not pose a genotoxic hazard and, therefore, should not be considered support for the classification of glyphosate as a genotoxic carcinogen. The assessment of the epidemiological data found that the data do not support a causal relationship between glyphosate exposure and non-Hodgkin's lymphoma while the data were judged to be too sparse to assess a potential relationship between glyphosate exposure and multiple myeloma. As a result, following the review of the totality of the evidence, the panels concluded that the data do not support IARC's conclusion that glyphosate is a 'probable human carcinogen' and, consistent with previous regulatory assessments, further concluded that glyphosate is unlikely to pose a carcinogenic risk to humans.

Other pro-industry articles include a paper by Boobis and co-authors (Boobis AR 2016) criticizing the agency's hazard identification approach as "outmoded." The publication maintains that the IARC classifications of carcinogenic to humans, probably carcinogenic to humans, or possibly carcinogenic to humans "can all lead to negative publicity and a 'health scare.'" In 2017 Bus submitted an article critiquing IARC's use of oxidative stress as one of the mechanistic tests for carcinogenicity (Bus JS 2017). While this was one of several genotoxic tests for glyphosate described in the IARC 2017 monograph, it was specifically identified by Smith and co-authors [26] as one group of the ten groups of mechanistic tests for carcinogenicity. Bus cites a number of issues which he believes demonstrate deficiencies, questioning the role that oxidative stress may play in carcinogenic development. In particular, Bus makes the following observation:

> a significant amount of caution must be raised as to the possibility that glyphosate MoA [mode of action] investigations represent an example of a mechanism(s) in search of an outcome, rather than the expected path of using mechanisms to ascertain the human relevance of cancer findings unequivocally demonstrated in animals

An opinion paper published by Robert Tarone in 2018 [27] presents a critical appraisal of several aspects of IARC's glyphosate deliberations. Included were IARC's assessment of animal studies. Tarone argues that data from the Greim publication [11] was "made available before the Monograph 112 Working Group meeting." In addition to his argument regarding the Greim publication, Tarone criticizes the Working Group's analyses of the epidemiologic studies, as well as the composition and professional affiliations of the IARC Working Group members.

A subsequent editorial piece by Tarone [28] repeats the critical comments of the earlier publication, and also notes the following:

> In particular, they [IARC and IARC staff] falsely claim that my paper was paid for by Monsanto. The basis for this claim appears to be a post on the IARC website noting my participation in a three hour meeting in September 2015 with a lawyer representing Monsanto to discuss the IARC Monographs Program review process....This meeting was known to IARC only because I acknowledged it in my published paper....No payment was received for writing the paper, nor was I requested to write the paper by anyone associated with Monsanto.[2]

It is noteworthy that, while not specifically mentioning the Tarone opinion piece in their article, Infante and co-authors [12] note that:

> Monsanto, through membership in the ACC [American Chemistry Council], has lobbied extensively, and paid scientists to author papers on the safety and continued use of glyphosate.

In contrast, beyond IARC's own response that was cited earlier, other members of the scientific community have submitted articles responding to publications critical of IARC. A publication by Myers et al. [22] provides a detailed description and consensus statement regarding concerns about the use of glyphosate and glyphosate-based herbicides. The introductory material summarizes a number of issues: (1) that glyphosate use has increased dramatically during the period from its introduction to 2016; (2) that residues of glyphosate can be found in food; (3) that data do show hazards associated with exposure (several are cited); (4) that weed tolerance has resulted in the use of herbicide mixtures as well as the introduction of genetically engineered plants that carry multiple herbicide-resistant genes; and (5) that regulatory authorities have not set consistent levels of acceptable daily intake for glyphosate in food products.

The paper provides multiple examples supporting each of the claims cited above and also lists uncertainties in the then-current assessment. These include:

- Lack of testing for herbicide resistance
- Lack of information about adjuvants
- A need for more testing at low exposure levels (not generally used in animal toxicology studies)
- Lack of large-scale biomonitoring studies that look at responses in the general population
- Lack of large-scale surveys of glyphosate and glyphosate/adjuvant contamination in food
- Lack of control for glyphosate contamination in food standardly fed to laboratory animals used for toxicity testing (which could influence responses in "control" animals).

Citing these concerns, the authors issued a consensus statement recommending the following:

- The need for regulatory tests conducted by those who are independent of scientists representing organizations that are seeking registration

---

[2] As noted in the previous chapter, Dr. Tarone testified before the Congressional Committee on Science, Space and Technology in 2018. His testimony questioned IARC's designation of glyphosate as a Category 2A carcinogen, and by extension, more stringent regulations on its use.

- The need for more epidemiologic studies
- The need for better biomonitoring studies such as the US CDC's National Health and Nutritional Examination Survey (NHANES) to establish whether glyphosate and/or AMPA accumulate in the body
- Studies should be conducted to establish whether those exposed to glyphosate early in life are more susceptible to effects from exposure to other chemicals or "other stressors"
- Studies are needed to assess whether glyphosate is an endocrine disruptor
- The need for an increase in the number of different species used in toxicology studies; studies should cover the entire lifespan of the species
- Tests conducted by NTP should look not only at glyphosate but also glyphosate with commonly used adjuvants.

Echoing parts of the consensus statement developed by Myers et al. [22], and countering some of the arguments against the IARC classification cited earlier, in 2018 Kawada published a short letter to the editor of Experimental and Clinical Sciences Journal [14]. In particular, the author noted that protective equipment such as that worn by participants in the AHS would lessen the likelihood of exposure and that more large-scale monitoring for human exposure, while challenging, would provide a better estimate of population exposure and cancer incidence. Citing a 2018 study by Connolly [6] Kawada notes that glyphosate was detectable in 20% of Irish adults with no occupational exposure to glyphosate. A study in China cited in an earlier chapter [30] noted that workers exposed occupationally to glyphosate during its production were found to have detectable urinary concentrations both of glyphosate and AMPA and that concentrations correlated with the time-weighted average glyphosate exposure.

10.1.3.2 The Commercial Sector, Litigation, and Coverage by the Lay Press

Since the IARC classification of glyphosate was announced, a number of articles have appeared in the lay press both criticizing and supporting IARC's carcinogenic hazard designation. An article by Geoffrey Kabat published in Forbes Magazine [13] entitled "IARC's Glyphosate-gate Scandal" took a basically negative tone toward the glyphosate designation. Kabat cites two issues of concern that were noted by Robert Tarone regarding animal and epidemiology studies: (1) the lack of clear dose–response relationships in the animal studies; and (2) the lack of inclusion of certain findings in the human epidemiology studies. Kabat's article also discusses the role of Christopher Portier in the IARC proceedings.

A Reuters article filed in 2017 by Kelland [15] maintains that changes were made between the IARC draft document, which, according to Kelland, did not conclude that glyphosate led to animal tumors and the final published document which concluded that it did. Specifically, the Reuters article cites testimony by Charles Jamison, the toxicologist who chaired the animal studies IARC subgroup that reviewed findings of glyphosate's effect on laboratory animals. According to Kelland, Jamison was said to have told Monsanto attorneys involved in a personal injury lawsuit against Monsanto, that "he did not know when, why or by whom the edits had been made."

Kelland wrote several other articles critical of IARC's work. Of particular interest is another [16] that focused on the epidemiology evidence considered by IARC. The article notes that:

> Previously unreported court documents reviewed by Reuters from an ongoing EU legal case against Monsanto show that Blair knew the unpublished research found no evidence of a link between glyphosate and cancer. In a sworn deposition given in March this year in connection with the case, Blair [Chair of the IARC Working Group] also said the data would have altered IARC's analysis. He said it would have made it less likely that glyphosate would meet the agency's criteria for being classed as 'probably carcinogenic.'

Carey Gillam[3] in an article published by Huffington Post [10] provides a rebuttal to the Reuters article [16] that implied Blair withheld publication of epidemiological findings that would have shown glyphosate was not linked to the development of non-Hodgkin's lymphoma. In thoroughly examining Blair's deposition, Gillam notes that the Reuters story selectively took information regarding one unpublished research study (in progress at the time of the IARC meeting). In other ways Blair, in his deposition, discussed studies that strongly related glyphosate exposure to cancer development. To underscore her point, Gillam provided a link to Blair's full deposition. In this same piece, Gillam also notes that Robert Tarone who is quoted in the Kabat submission [13] is a paid consultant of Monsanto.

At the same time, lay publications and, not surprisingly, IARC's classification of glyphosate have led to significant litigation which continues. Given the different nature of various lawsuits (individual litigants, class action suits, etc.), it is difficult to document the precise amount of litigation, although the growth has been rapid. To illustrate, Krimsky and Gillam [19] report that "as of November 2017, roughly 3500 plaintiffs had cases ongoing against Monsanto." In March 2020, New York Times reporter Patricia Cohen reported that Bayer, the German conglomerate that purchased Monsanto in 2018, had put aside ten billion dollars as a settlement to cover both the estimated 95,000 cases already litigated as well as estimates of the number of future lawsuits brought against Bayer. The New York Times article also noted that the settlement would not cover an estimated 30,000 claims by litigants who did not agree to take part in the settlement. Interestingly, the article pointed out that Bayer is "continuing to sell the product without adding warning labels about its safety." In June of 2020, the law firm of Weitz and Luxenberg reported that they and other law firms had reached settlements "with Monsanto" for almost 100,000 Americans who had developed cancer (Weitz and Luxenberg 2020).

Relatively early following the publication of IARC Monograph 112, as a result of litigation filed in California, Judge Vince Chhabria, a federal judge for the US District Court for the Northern District of California, ruled that the documents plaintiffs were able to acquire in litigation against Monsanto Roundup were to be unsealed.

[3] Carey Gillam, presently research director for US Right to Know, was for a number of years an employee of Reuters

Generally referred to as the "Monsanto Papers," these documents provide information about the methods used by Monsanto to misrepresent research findings on glyphosate and Roundup™.

Although Monsanto initially made efforts to prevent the release of the internal documents, ultimately Monsanto did release the papers, albeit gradually over time. In 2019 the law firm of Baum, Hedlund, Aristei, Goldman, the firm that originally requested the release of the Monsanto documents, published a report with links to a 118-page spreadsheet listing what had been made available at that time [2]. Among the documents released by Baum et al. [2] are those that provide information about a relationship between Monsanto and Kate Kelland, the reporter for Reuters who claimed that Charles Jamison (the toxicologist who chaired the animal studies IARC sub-group reviewing findings of glyphosate's effect on laboratory animals) testified that changes that had been made between the draft report on animal carcinogenicity and the final report. The original document [1] includes numerous quotes from Dr. Jamison that do not include any reference to the original draft statement or any "alterations." It does, however, provide a concise summary of his appraisal:

> Considering all data from the mice and rat studies I reviewed, there is "Sufficient" evidence that shows glyphosate is carcinogenic in experimental animals causing kidney tumors, hemangiosarcomas, malignant lymphoma, adenocarcinomas of the lung, and hemangiomas in mice and pancreatic cell tumors, interstitial cell tumors in the testes, hepatocellular adenomas, and thyroid follicular cell tumors in rats. This statement is based on my stated criteria of a causal relationship between exposure to glyphosate and an increased incidence of malignant and/or a combination of malignant and benign tumors, in multiple species, at multiple tissue sites, from multiple studies, and to an unusual degree with regard to incidence, site, or type of tumor.

Of particular interest in the Monsanto Papers are the series of articles cited in Critical Reviews in Toxicology that were discussed earlier. When published, the editor of the series noted that "Neither any Monsanto company employees nor any attorneys reviewed any of the Expert Panel's manuscripts prior to submission to the journal" [20]. However, the unsealed documents suggest that Monsanto scientists were heavily involved in organizing, reviewing, and editing drafts submitted by the outside experts. A spokeswoman from Taylor & Francis, publisher of Critical Reviews in Toxicology, noted that an investigation is underway. In October 2017, scientists at the Center for Biological Diversity, Center for Food Safety, Pesticide Action Network, and Center for Environmental Health called for a retraction of one of the series reviews. The group noted that "These are serious offenses and if left unanswered will ultimately undermine the work of many scientists who view scientific ethics to be sacrosanct" [2].

A 2018 publication by McHenry provides more specific information about the categories of "misrepresentation of scientific testing" used by Monsanto [25]. As well as ghost-writing, discussed earlier, Monsanto both interfered with the peer review process and worked to retract a paper published by Séralini et al. [24]. The publication reports an increase in tumors among rats fed genetically modified corn and Roundup™. Even though the paper was published, Monsanto played an important role in facilitating "numerous third-party expert letters" to the editor of Food and

Chemical Toxicology (which had published the Séralini report). The solicited letters were published in a later edition of the same journal. Certain of these letters described numerous significant deficiencies, poor study design, and selective statistics that the writers purported to find in the Séralini paper [21]. Of special importance was the relationship between A. Wallace Hayes, the Editor-in-Chief of Food and Chemical Toxicology and Monsanto, as well as the role that Monsanto played in contacting scientific experts encouraging them to develop their letters to the editor. Ultimately Food and Chemical Toxicology retracted the paper, citing that the published findings were "inconclusive," and, although it was published elsewhere [24], the IARC Working Group did not consider the findings in its deliberations [23].

McHenry documents additional relationships between Monsanto and editors of other journals, as well as a "hidden" role in "Academics Review: Testing Popular Claims Against Peer Reviewed Science." Although this publication purports to be a product of an independent non-profit organization involving professors of "food-related microbiology, nutritional, and safety issues," it has attacked the 2017 IARC Monograph and several journalists. Internal Monsanto documents indicate that funds were transferred to a third party so that experts could be compensated for providing responses. One of the sources of this funding was the Council for Biotechnology Information of which a senior member of Monsanto was a member of the Council for Biotechnology Information board of directors.

References

1. Baum H, Aristei G (2017) Expert report of Dr. Charles W. Jameson, Ph.D. on Behalf of Plaintiffs. https://www.baumhedlundlaw.com/documents/pdf/monsanto-documents/charles-jameson-expert-report.pdf
2. Baum H, Aristei G (2020) Monsanto Papers Secret Documents. H. Baum, Aristei, Goldman. Los Angeles, CA. https://www.baumhedlundlaw.com/toxic-tort-law/monsanto-roundup-lawsuit/monsanto-secret-documents/
3. Becker RA, Dreier D, Manibusan MK, Cox LA, Simon TW, Bus JS (2017) How well can carcinogenicity be predicted by high throughput "characteristics of carcinogens" mechanistic data? Regul Toxicol Pharmacol 90:185–196. DOI: https://doi.org/10.1016/j.yrtph.2017.08.021
4. Bobbis AR, Cohen SM, Dellarco VL, Doe JE, Fenner-Crisp PA, Moretto A, Pastoor TP, Schoeny RS, Seed JG, Wolf DC (2016) Classification schemes for carcinogenicity based on hazard-identification have become outmoded and serve neither science nor society. Regul Toxicol Pharmacol 82:158–166. DOI: https://doi.org/10.1016/j.yrtph.2016.10.014
5. Chiu WA, Guyton K, Martin MT, Reif DM, Rusyn I (2018) Use of high-throughput in vitro toxicity screening data in cancer hazard evaluations by IARC monograph working groups. ALTEX 35(1):51–64. https://doi.org/10.14573/altex.1703231
6. Connolly A, Leahy M, Jones K, Kenny L, Coggins MA (2018) Glyphosate in Irish adults - a pilot study in 2017. Environ Res 165:235–236. DOI: https://doi.org/10.1016/j.envres.2018.04.025
7. Corporate Observatory Europe (2017) Setting the record straight on false accusations. Dr C. Portier's work on glyphosate and IARC. https://corporateeurope.org/en/food-and-agriculture/2017/10/setting-record-straight-false-accusations-dr-c-portier-work-glyphosate

8. Director, IARC, (2018) IARC response to criticisms of the monographs and the glyphosate evaluation. https://www.iarc.who.int/wp-content/uploads/2018/07/IARC_response_to_criticisms_of_the_Monographs_and_the_glyphosate_evaluation.pdf
9. Escher BI, Henneberger L, Konig M, Schlichting R, Fischer FC (2020) Cytotoxicity burst? Differentiating specific from nonspecific effects in tox21 in vitro reporter gene assays. Env Health Perspt 128(7):077007. DOI: https://doi.org/10.1289/EHP6664
10. Gillam C (2017) Monsanto spin doctors target cancer scientist in flawed reuters story. Huffington Post. New York. https://careygillam.com/articles/article/monsanto-spin-doctors-target-cancer-scientist-in-flawed-reuters-story
11. Greim H, Saltmiras D, Mostert V, Strupp C (2015) Evaluation of carcinogenic potential of the herbicide glyphosate, drawing on tumor incidence data from fourteen chronic/carcinogenicity rodent studies. Crit Rev Toxicol 45(3):185–208. https://doi.org/10.3109/10408444.2014.1003423
12. Infante PF, Melnick R, Huff J, Valnio H (2018) Commentary: IARC Monographs Program and public health under siege by corporate interests. Am J Ind Med 61:277–281. DOI: https://doi.org/10.1002/ajim.22811
13. Kabat G (2017) IARC's Glyphosate-gate Scandal. Forbes. Jersey City, NJ, Forbes. https://www.forbes.com/sites/geoffreykabat/2017/10/23/iarcs-glyphosate-gate-scandal/?sh=6330efd91abd
14. Kawada T (2018) Glyphosate toxicity and carcinogenicity. EXCLI J 17:800–801. https://pubmed.ncbi.nlm.nih.gov/30233278/
15. Kelland K (2017) Glyphosate battle. Reuters investigates. London, Reuters
16. Kelland K (2017) Cancer agency left in the dark over glyphosate evidence. Reuters investigates. London, Reuters. https://www.reuters.com/investigates/special-report/glyphosate-cancer-data/
17. Kier LD, Kirkland DJ (2013) Review of genotoxicity studies of glyphosate and glyphosate-based formulations. Crit Rev Toxicol 43(4):283–315. DOI: https://doi.org/10.3109/10408444.2013.770820
18. Kimmel GL, Kimmel C, Williams AL, DeSesso JM (2013) Evaluation of developmental toxicity studies of glyphosate with attention to cardiovascular development. Crit Rev Toxicol 43(2):79–95. DOI: https://doi.org/10.3109/10408444.2012.749834
19. Krimsky S, Gillam C (2018) Roundup litigation discovery documents: implications for public health and journal ethics J Public Health Policy 39(4). https://medicine.tufts.edu/sites/default/files/pub2018RoundupLitigationDiscovery.pdf
20. McClellan RO (2016) Evaluating the potential carcinogenic hazard of glyphosate. Crit Rev Toxicol 46. https://doi.org/10.1080/10408444.2016.1234117
21. McHenry LB (2018) The Monsanto papers: poisoning the scientific well. Int J Risk Safety Med 29:193–205. DOI: https://doi.org/10.3233/JRS-180028
22. Myers JP, Antoniou M, Blumberg B, Carroll L et al (2016) Concerns over use of glyphosate-based herbicides and risks associated with exposures: a consensus statement. Env Health 15(19). https://ehjournal.biomedcentral.com/articles/10.1186/s12940-016-0117-0
23. Richmond ME (2018) Glyphosate: a review of its global use, environmental impact, and potential health effects on humans and other species. J Environ Stud Sci 8(4):416–434. https://doi.org/10.1007/s13412-018-0517-2
24. Séralini GE, Clair E, Mesnage R, Gress S, Defarge N, Malatesta M, Hennequin D, de Vendômois JS (2014) Republished study: long-term toxicity of a roundup herbicide and a roundup-tolerant genetically modified maize. Environ Sci Eur 26(1)
25. Séralini GE, Clair E, Mesnage R, Gress S, Defarge N, Malatesta M, Hennequin D, de Vendomois JS (2012) Long term toxicity of a roundup herbicide and a roundup-tolerant genetically modified maize. Food Chem Toxicol 50:4221–4231. doi: https://doi.org/10.1016/j.fct.2012.08.005
26. Smith MT, Guyton K, Gibbons CF, Fritz JM, Portier CJ, Rusyn I, DeMarini DM, Caldwell JC, Kavlock RJ, Lambert PF, Hecht SS, Bucher JR, Steward BW, Baan RA, Cogliano VJ, Straif K (2016) Key characteristics of carcinogens as a basis for organizing data on mechanisms of carcinogenesis. Environ Health Perspect 124(6):713–721. https://doi.org/10.1289/ehp.1509912

27. Tarone RE (2018) Conflicts of interest, bias, and the IARC monographs program. Reg Tox and Pharmacol 98:A1–A4. DOI: https://doi.org/10.1016/j.yrtph.2018.09.005
28. Tarone RE (2018) Comments on the IARC classificaetion of glyphosate as a probable human carcinogen. Comments to the hearing before the committee on science, space, and technology, House of Representatives, one hundred and fifteenth congress, second session. Serial No. 115–46. https://www.govinfo.gov/content/pkg/CHRG-115hhrg28933/pdf/CHRG-115hhrg28933.pdf
29. Williams GM, Aardema M, Acquavella J, Berry C, Brusick D, Burns MM, Camargo JLV et al (2016) A review of the carcinogenic potential of glyphosate by four independent expert panels and comparison to the IARC assessment. Crit Rev Toxicol 46:3–20. https://doi.org/10.1080/10408444.2016.1214677
30. Zhang F, Xu Y, Liu X, Pan L, Ding E, Dou J, Zhu B (2020) Concentration distribution and analysis of urinary glyphosate and its metabolites in occupationally exposed workers in Eastern China. Int J Environ Res Public Health 17(8):2943. DOI: https://doi.org/10.3390/ijerph17082943
31. Taylor E (2020) Law firm Weitz & Luxenberg says has reached Glyphosate settlement with Bayer. https://www.Edward.Taylor@thomsonreuters.com
32. California Office of Environmental Health Hazard Assessment (2017). Glyphosate listed effective July 7, 2017, as Known to the State of California to cause Cancer

Part IV
Environmental Justice Issues: Domestically and Globally

Chapter 11
Domestic Environmental Justice Considerations

Abstract Data from carefully executed human exposure studies, while an essential part of the information that IARC used to assess cancer hazards of the pesticides and herbicide evaluated in Monograph 112, may be only the "tip of the iceberg" in terms of hazards posed to the large number of agricultural and nursery workers, who are employed both in developed and developing countries. These workers are exposed on an almost daily basis and over long periods of time. It is essential, when considering exposure impact, that the health and safety of these workers be taken into account. Whether domestically or globally, such individuals often represent underserved constituencies. This underscores questions of environmental justice, issues that are critical in considering the overall impact of hazard identification. For multiple reasons: the migratory and seasonal work of many farm and nursery workers; the lack of adequate personal protection; the quality of living conditions and lack of good sanitation; language barriers; the lack of good health-care facilities where long-term outcomes can be documented; the lack of health care providers who are aware of the spectrum of risks and symptoms from pesticide and herbicide exposures; and an uneven regulatory network, such workers frequently are poorly protected. The following chapter focuses on these questions in the US. While the discussion is confined to domestic issues, on a global basis, many, if not most of these questions also apply.

11.1 Evolution of Steps Intended to Address Environmental Justice

The etiology of the environmental justice movement in the US is complex, having grown from a number of different situations and from multiple causes. It is helpful to consider how a number of events came together that, in theory, gave rise to regulations designed to provide remedies.

In its 2016 publication, the National Resources Defense Council (NRDC) describes several major events that gave birth to the movement. Perhaps the most seminal of these were the 1982 protests and litigation filed by residents of Warren

M. Richmond, *Cancer Hazards: Parathion, Malathion, Diazinon, Tetrachlorvinphos and Glyphosate*, AESS Interdisciplinary Environmental Studies and Sciences Series,
https://doi.org/10.1007/978-3-030-81953-8_11

County, NC who were responding to the state's decision to dump truckloads of polychlorinated biphenyl (PCB)-contaminated soil in a landfill in Afton, a small community in Warren County. The county is rural, and its residents were socio-economically disadvantaged and predominantly black. While ultimately, following litigation, the state of North Carolina prevailed, the publicity that the demonstrations and litigation received was effective in highlighting the environmental injustice experienced by residents [17].

The Afton demonstrations and actions were one of a number of movements protesting environmental injustices that predominantly affect disadvantaged communities, often people of color. Important earlier events include the protests by Latino farm workers working in San Joaquin Valley, CA that were organized by Cezar Chavez; demonstrations by African- Americans in Houston opposing a garbage dump that had resulted in the loss of two children's lives; and opposition to citing of a sewage treatment plant by residents of East Harlem in New York City. However, according to NRDC "the Warren County protests marked the first instance of an environmental protest by people of color that garnered widespread national attention" [17].

In the aftermath of the Warren County protests, a number of church groups and governmental bodies looked more closely at the questions encompassing environmental justice. In 1991 the National People of Color Environmental Leadership Summit met in Washington, D.C., and early in the Clinton Administration Executive Order 12898 was issued. The intent of the executive order was to address environmental justice in all its aspects.

Executive Order 12898 issued by President Clinton in 1994 required:

> federal agencies to: identify and address the disproportionately high and adverse human health or environmental effects of their actions on minority and low-income populations, to the greatest extent practicable and permitted by law; to develop a strategy for implementing environmental justice; to promote nondiscrimination in federal programs that affect human health and the environment; to provide minority and low-income communities access to public information and public participation [15].

While the order establishes steps for addressing environmental justice issues, and, theoretically requires mitigation measures, it does not specifically define the term environmental justice, nor is it specific in terms of "environmental justice review" or the establishment of appropriate mitigation measures. Given these limitations, addressing environmental justice has, not surprisingly, become an issue that requires specific definitions, specific "environmental justice review" processes and mitigation that are unique to different situations. To some extent, EPA has broadened these issues in terms of a number of directives. The EPA legal definition of environmental justice, as stated in 2020, specifies that:

> Environmental Justice is the fair treatment and meaningful involvement of all people regardless or race, ethnicity, income, national origin or educational level with respect to the development, implementation and enforcement of environmental laws, regulations and policies. This goal will be achieved when everyone enjoys:

- the same degree of protection from environmental and health hazards, and

- equal access to the decision-making process to have a healthy environment in which to live, learn, and work [7].

Beyond both the Executive Order and EPA's work, the movement has developed and evolved over a number of years. Ironically, this has led to concerns regarding the ways that environmental justice has been addressed which has had different degrees of focus and levels of success. This was discussed in some detail by Gauna [11] citing both the broad field of issues that fall under the umbrella of environmental justice as well as the types of attention given to specific areas. While the specific details of legal attention to environmental justice issues as a whole are likely to differ from 2002, when the article was published, as compared to today, the proportion of all legal publications focusing on farmworker environmental justice in 2002 was "less than a handful" of the total legal articles published at that time, and this has not changed significantly. Guana attributes this to the EPA regulatory structure designed to address farmworker protections, specifically, the Federal Insecticide Fungicide and Rodenticide Act (FIFRA) which became incorporated into EPA. The provisions of FIFRA "presented EPA with a difficult benefit–cost issue." Subjected to these criteria, it has become, at best, challenging to address environmental justice issues for farmworkers.

The lack of focus on environmental justice as it applies to farm and nursery workers can also be attributed to other factors. One issue may be defining "environmental justice" so that the definition adequately captures specific characteristics and needs of farm, nursery, and landscape workers. Numerous definitions and understandings of environmental justice exist. Some, as in the case of the North Carolina community, address environmental justice issues affecting those in a given location. Others deal with aspects of occupational issues such as the manufacturing or meat-packing sector. Few consider cohorts that may move from location to location rather than working continuously in a defined location. This is frequently true with many farm, nursery, and landscape workers.

For purposes of the following discussion, I will use a productive definition as characterized by Berkey which does incorporate these issues. Berkey's designation defines environmental justice as "A form of justice based on addressing the political-economic structures that produce environmental problems, aimed at creating a system within which we focus on causes rather than symptoms. Emphasizes participation in the decisions through which environmental burdens are produced. Characterized by a movement from 'not in my backyard' to a 'not in anyone's back yard political frame" [4]. Key to this definition is that farm, landscape and nursery workers are a community defined by a shared occupation despite such issues as migratory patterns or different types of crops. Such communities share an occupation but may or may not share a geographic location.

Another challenge to addressing environmental justice in the farming, landscaping and nursery workers is the concept of "agricultural exceptionalism." Agricultural exceptionalism originated during the Roosevelt administration; a measure that was adopted by the administration as a compromise with Southern Democrats. Farmworkers would be excluded from protections offered in the National Labor Relations

Act of 1935 in order to win votes of Southern Democrats for supporting other labor groups. At the time that these protections were established, a large number of farmers in the South were people of color.

11.2 Specific Challenges Farm, Nursery, and Landscape Workers

Even within the occupational sector, these workers present unique challenges, and not surprisingly, these challenges have been the subject of a number of investigations. Arcury and co-investigators [3], examining the perception of pesticide risk and the sense of control over pesticide hazards, conclude that the two are not closely connected. They note that "[a] key tenet of environmental justice is that communities must have control over their environment." They argue that for "pesticide safety education to be effective, it must address issues of farmworker control in implementing workplace pesticide safety."

In 2004 The Farmworker Pesticide Exposure Comparable Data Conference was convened to develop consensus regarding farmworker pesticide exposures [2]. The conference identified four broad areas requiring attention: exposure assessment from the environment; biomarkers of exposure; predictors of exposure (personal and occupational); and health outcomes. Within these areas a number of challenges were characterized. These include:

1. The number of farmworkers at risk. Given the uncertainties of present policies regarding immigration or guest workers, the exact number of farmworkers/nursery workers is difficult to estimate. One recent estimate is approximately 800,000 employed workers and about half undocumented.[1] Other sources consider the number to be closer to be between two and one half and three million with the percentage of undocumented workers to be roughly the same [8].
   The make-up of workers is diverse. Findings from the most recent US Department of Labor National Agricultural Workers Survey [12] which summarized 2015-16 findings reported the following:
   a. 69% of workers were born in Mexico
   b. 24% were born in the US
   c. Approximately 6% were born in Central America
   d. 1% came from other areas
   e. 51% of all workers had authorization; of these 29% were US citizens, 21% permanent residents and 1% had some other type of authorization.
2. Mobility of workers. While some workers may move from place to place, this number is thought to be decreasing. A US Department of Labor report

[1] Personal communication, Anne Katten, Director, Pesticide and Worker Safety Project, California Rural Legal Assistance Foundation.

published in 2018 [12] indicated that a number of farmworkers worked at only one facility during the years 2015-2016. However, given the more recent immigration policies, it may be difficult to specify the present degree of mobility.

3. Residence status. As noted earlier [16], US citizens or permanent residents, whether mobile or not, may be "more secure," and therefore more likely to return to or remain at specific places of work. Holders of H-2A visas, because they are hired for specific work, are less mobile. While, in the past, undocumented workers may have chosen to move more frequently, depending on such issues as needs for crop-harvesting which can be a form of "protection," if they fear deportation, they may also now be less likely to be mobile.
4. Obstacles to communication. Several factors contribute to these obstacles.

    a. Language. Although the largest number of farm/landscape workers who do not speak English are Hispanic, they may speak one of several dialects or use words unique to their region. Additionally, a number of workers coming from Mexico or Central America represent indigenous cultures and communicate in their own languages.
    b. Education and literacy. Hand-in-hand with spoken language challenges are challenges in literacy. According to the US Department of Labor Survey published in 2018 [12], the average educational attainment of farm, nursery, and landscape workers is at the 8th grade level; 4% have no education, and 37% have an education at the 6th grade level or lower. Reading in their own language may be difficult. In the case of those representing indigenous cultures, the language may not be recorded. Even when recorded, it may be difficult to find local experts who can speak with or prepare signs and other forms of information that can be understood. According to Anne Katten of the California Rural Legal Assistance Foundation, in California efforts have been made to prepare signs with drawings to better explain health and safety issues (see footnote 1).

5. Exposure assessment and biomonitoring. Although surveillance data provide a relatively rudimentary measure of exposure, to measure individual exposures and conduct biomonitoring, individual samples of urine, blood or both must be collected. This first requires a sampling strategy that identifies various points of access to farmworkers, including worker camps, individual residences, clinics or community organizations. As noted by Arcury [2], each of these venues may have its own "gatekeeper who controls access, and each gatekeeper may have reasons to restrict research" for multiple reasons ranging from concerns about productivity to issues of grower liability. In addition, workers themselves may not wish to donate samples, fearing retribution from various interest groups. Finally, all samples must be properly stored for analysis, and some health facilities are not equipped for significant sample storage.
6. Health outcomes. It is often difficult to collect information on short-term health outcomes and to relate these to exposure data. As noted by Berkey [4], many

workers, fearing retribution, are hesitant to seek health care. Health care facilities, with rare exceptions, are not equipped to diagnose long-term chronic conditions, in part, because they may not have the means to follow individuals over long periods of time. Furthermore, as noted by Kelley [13], health care practitioners have limited knowledge about agricultural work, and may not notice health end-points that are associated with this type of work. Such end-points may not be specific to pesticide or herbicide exposures.

Additional concerns relate to living conditions and access to adequate health care. As noted by Flocks [10], many farmworkers live in "unsanitary and overcrowded housing." Lacking access to proper sanitation can lead to inadvertent pesticide exposure. In addition, even if covered by some type of financial assistance, health care may not be readily available for many reasons: communication challenges (as described above), inability to get time away from work, insufficient funds either to cover gaps in possible medical coverage or lack of knowledge about coverage. Finally, because workers fear deportation, many may not seek medical care, even when measures are in place for their protection.

Women farm and nursery workers present an especial challenge. Many are of childbearing age, live with a spouse or partner, and become pregnant while working. A small-scale community-based participation research (CBPR) study of nursery and fern workers conducted by Flocks and co-investigators [9] found that fernery and nursery workers, who work either in enclosed spaces or work in close full body contact with ferns that may have pesticide residue, described a number of health problems, some of which might affect pregnancy or have a negative effect on the developing fetus. The workers participating in the study did not point out their own fears about a negative pregnancy outcome, but cited information about adverse outcomes among other workers. Another article published by Flocks in 2012 [10] includes details of several infants born with severe birth defects to farmworkers in southwest Florida. These births, in 2004–2005, and the associated exposures of the mothers to several pesticides during the early period of gestation, were investigated by a number of agencies. Findings indicated that the mothers had entered fields prior to expiration of the re-entry period following pesticide application; however, no agency could establish any causal relationship between the pesticides and birth defects. This was reported in some detail by Calvert and co-authors [6], and raised a number of questions about poor regulation of pesticide exposures, irrespective of the causal association with negative birth outcomes.

Perhaps the most comprehensive series of studies looking at effects of farmworker women and their children is the Center for the Health Assessment of Mothers and Children of Salinas (CHAMACOS) Study. The study is directed by Dr. Brenda Eskenazi, and is affiliated with the University of California at Berkeley. Interestingly, the study acronym CHAMACOS is also Mexican Spanish for little children. Since 1999 the study has followed children born to farmworker women who live in the Salinas valley and who have been exposed in utero and following birth to a number of pesticides sprayed onto the farming fields of the valley. Presently, approximately 150 publications have come from the research. Significant numbers of these publications focus either on outcomes from exposure to organophosphate pesticides, with end points either representing aspects of neurodevelopment or of respiratory development.

Studies focusing on neurodevelopment end points include those of Bouchard and co-workers cited earlier [5], regarding the inverse relationship between deficits in IQ points and the presence of urinary metabolites of organophosphates in maternal urine during the gestation period, as well as decreases in pulmonary function correlated with organophosphate exposure among children of farmworkers [16]. Pulmonary function decreases were better correlated with biomarkers of exposure in children.

11.3 Regulatory and Other Protections for Agricultural/Nursery Workers

A chapter by Arcury and Quant in their book "Latinx Farmworkers in the Eastern United States," [1] notes that, among other issues that complicate the use of measures to address farmworker environmental health and justice, deficits result "largely" from agricultural labor policy. Despite the attention that has been given to farm, landscape, and nursery workers in the past two decades, the history of protections has limited the effectiveness of regulations.

As noted by Flocks [10], "successful protective regulation requires that government enforce standards, that management complies with standards, and that workers are knowledgeable about standards." For farmworkers this should involve both standards applying to exposure conditions, and measures to ensure workers can control their working conditions. That strong measures are not in place for farm and nursery workers derives in large part from the concept of "agricultural exceptionalism" discussed earlier in this chapter. Today this continues to mean that farm, landscape, and nursery workers are protected by a number of regulations, both state and federal, and that elements of this "patchwork" of regulations may at times work against each other.

Perhaps the first protective regulation for farm and nursery workers came in 1947 when the Federal Insecticide, Fungicide and Rodenticide Act (FIFRA) was passed by Congress. Today FIFRA is part of EPA and is enforced under EPA regulations. The EPA regulations as they apply to farm, landscape, and nursery workers do specify the re-entry period following treatment with pesticides and herbicides. However, these provisions are often circumvented by such practices as using labor contractors or temporary workforce personnel (Flocks JD 2012b). While the Occupational Safety and Health Administration (OSHA) should have some regulatory authority over farm and nursery worker safety and health, the administration of FIFRA under EPA and a prohibition clause in the act establishing OSHA has made it impossible for the Labor Department to use OSHA to enforce FIFRA provisions.

The Worker Protection Standard (WPS), also administered under EPA, specifies specific guidelines for protecting workers. Included are the following:

1. Worker and handler training
2. Hazard communication
3. Pesticide safety information
4. Notification [identification] of treated areas
5. Minimum worker age
6. Re-entry restrictions
7. Exemptions and exceptions to requirements
8. Personal protective equipment requirements
9. Provision of decontamination supplies
10. Establishment of emergency assistance.

While, ideally, the WPS provisions would afford workers significant protection, the rules may not be followed (as for example, in the case of Florida farm workers cited earlier). Hence, it remains challenging to provide adequate protection. It should be pointed out that the WPS, since its initial draft, has been updated. However, the updates required approximately 20 years, and many of the updates have not yet been implemented.

Focusing on these issues in Florida, one of the major growing areas in the US, Murphy-Greene and Leip [14] trace the development of regulatory approaches from the EPA and OSHA regulations in the 1960s and 1970s to the regulatory framework in 2002, and assess how effective this framework of approaches has been in adequately protecting workers in the state. Their evaluation was specifically looking at the success of Executive Order 12898 to protect farmworkers, while, at the same time, considering how specific measures should work in tandem with the executive order. In conducting their review, the authors list not only the federal agencies charged with farm and nursery worker protection, but also the Florida agencies and laws designed as well to address the same issues. They note that:

> Based on the various federal and state laws and the infrastructure of the agencies that are responsible for implementing environmental pesticide laws, it appears that there are good intentions for protecting farmworkers. However, the complexity of the laws and the disjointed nature of the federal and state agencies may well lead to implementation problems.

The authors specifically concluded that "The findings reveal that protective laws aimed at reducing environmental hazards for farm workers in South Florida are not being effectively implemented in the State of Florida" and that the goals of Executive order 12898 are "not being achieved".

While the doctrine of agricultural exceptionalism may have, on a national level, made it impossible to establish unions for farm and nursery workers, legislation in states and given locales has made the formation of advocacy groups possible. Nonetheless, on a national level, unions and other advocacy groups represent a small portion of the farm and nursery workers. In 2019, three states, California, Oregon and North Carolina had unions: in California, the United Farmworkers Union (UFW); in Oregon Pineros y Campesinos Unidos del Noroeste (PCUN); and in North Carolina the Farm Labor Organizing Committee (FLOC). Washington and New York have been undertaking efforts to organize more local advocacy groups. A smaller group,

Familias Unidas por la Justicia, has been organized, and efforts continue to organize farm workers in New York State.

Perhaps, because of the Agriculture Labor Relations Act passed by California in 1975, the UFW has had a strong voice in some circumstances. For example, Proposition 65, which required carcinogen warnings for Roundup, was passed. Although not providing as much protection as it might, Proposition 65 remains in effect today. At the same time, the UFW membership is made up primarily of workers on large-scale farms. In small-scale operations such as nurseries, it is difficult to connect with UFW and to establish union-negotiated worker protections.

Much remains to be done.

References

1. Arcury TA, Quandt S, "Latinx farmworkers in the Eastern United States." Concluding chapter "Conclusions: an updated agenda for farmworker social justice in the Eastern United States." pp. 253-266. Springer, 2020
2. Arcury TA, Quandt S, Barr DB, Hoppin JA, McCauley L, Grzysacz JG, Robson MG (2006) Farmworker exposure to pesticides: methodologic issues for the collection of comparable data. Environ Health Perspect 114:923–928. DOI: https://doi.org/10.1289/chp.8531
3. Arcury TA, Quandt S, Russell GB (2002) Pesticide safety among farmworkers: perceived risk and perceived control as factors reflecting environmental justice. Environ Health Perspect 110:233–240. DOI: https://doi.org/10.1289/ehp.02110s2233
4. Berkey B (2017) "Environmental justice and farm labor." Routledge, Taylor and Francis Group
5. Bouchard MF, Chevrier J, Harley KG, Kogut K, Vedar M, Calderon N, Trujillo C, Johnson C, Bradman A, Barr DB, Eskenazi B (2011) Prenatal exposure to organophosphate pesticides and IQ in 7-year-old children. Environ Health Perspect 119(8):1189–1195. DOI: https://doi.org/10.1289/ehp.1003185
6. Calvert GM, Alarcon W, Chelminski A, Crowley MS, et al. (2007) Case report: three farmworkers who gave birth to infants with birth defects closely grouped in time and place—Florida and North Carolina, 2004–2005. Environ Health Perspect 115(5):787–791. DOI: https://doi.org/10.1289/ehp.9647
7. EPA (2020) "Environmental Justice". https://www.epa.gov/environmentaljustice
8. National Center for Farmworker Health, Inc. (2020) "Facts about agricultural workers." http://www.ncfh.org/facts-about-agricultural-workers.html
9. Flocks J, Kelley M, Economos J, McCauley (2012a) Female farmworkers' perceptions of pesticide exposure and pregnancy health. J Immigrant Minority Health 14(4)
10. Flocks JD (2012b) The environmental and social injustice of farmworker pesticide exposure. Georg J Poverty Law Policy 19:255–282
11. Gauna E (2002) Farmworkers as an environmental justice issue: similarities and differences. Environs 25(2):67–76
12. JBS International (2018) "Findings from the national agricultural workers survey (NAWS) 2015–2016. A demographic and employment profile of United States Farmworkers." prepared for the U.S. Department of Labor, Employment and Training Administration, Office of Policy Development and Research by JBS International, Inc. https://www.dol.gov/sites/dolgov/files/ETA/naws/pdfs/NAWS_Research_Report_13.pdf
13. Kelley MA, Flocks J, Economos J, McCauley LA (2013) Female farm-workers' health during pregnancy: health care providers' perspectives. Work Health Saf 61(17):308–313. DOI: https://doi.org/10.1177/216507991306100706

14. Murphy-Greene C, Leip L (2002) Assessing the effectiveness of executive order 12898: environmental justice for all? Public Adm Rev 62(6):679–687. https://doi.org/10.1111/1540-6210.00250
15. EPA (2020) "Summary of Executive Order 12898 - Federal Actions to Address Environmental Justice in Minority Populations and Low-Income Populations." https://www.epa.gov/laws-regulations/summary-executive-order-12898-federal-actions-address-environmental-justice
16. Richmond M (2018) Glyphosate: a review of its global use, environmental impact, and potential health effects on humans and other species. J Environ Stud Sci 8(4):416–434. https://doi.org/10.1007/s13412-018-0517-2
17. Skelton R, Moller V (2016) The environmental justice movement. https://www.nrdc.org/stories/environmental-justice-movement

Chapter 12
The Global Importance of the IARC Hazard Identification

Abstract The issues of environmental justice associated with pesticide and herbicide exposure among farm and nursery workers that were discussed in the previous chapter are often far greater challenges in less developed countries. As food and flower production has become more global and the demand for exports and imports has grown, large-scale farming and nursery work in underdeveloped countries has also increased. Pesticide use has also become more widespread. This may include the use of pesticides and herbicides banned or restricted in developed countries that are often not subject to the same restrictions elsewhere. Agrichemical companies may choose to "dump" these chemicals, or, at times, they may continue to manufacture them for distribution in countries where they are poorly regulated or unregulated altogether. While pesticides and herbicides are not used exclusively in large-scale farming, the growth in large-scale farming implies more frequent availability and use in less developed countries. It is also more likely, for a variety of reasons, that these pesticides and herbicides may be used improperly, and that farmers and nursery workers are less likely to be protected. The overall effect of the growth of global production taken together with the higher volume of pesticide and herbicide use (often of banned substances) has created what is commonly referred to as the "circle of poison." The following chapter explores these issues and considers ways that they might be effectively addressed, including the importance of provisions that are being taken by IARC to enhance its work on cancer prevention

12.1 Overview

As noted in a recent UN document, the questions of hunger and food security on a global basis are frequently used as arguments for reconfiguring the processes of production, moving from small-scale farming to intensive agriculture that employs large volumes of pesticides and herbicides, theoretically to increase yields [20]. However, as noted in the same document "increased food production has not succeeded in eliminating hunger worldwide" [20]. In fact, it may have introduced an additional number of global health challenges.

M. Richmond, *Cancer Hazards: Parathion, Malathion, Diazinon, Tetrachlorvinphos and Glyphosate*, AESS Interdisciplinary Environmental Studies and Sciences Series,
https://doi.org/10.1007/978-3-030-81953-8_12

Whether nationally or globally, the processes of identifying and using optimal approaches to address food security and sustainable production frequently involve controversies between the industrial agriculture sector and proponents of a more holistic approach. It is likely that advocates for each approach will continue to strongly voice their views. At the same time, a global concern is growing about the increasing prevalence of fertilizers, pesticides, and herbicides in the environment and the toxic effects of this prevalence on human, plant, and animal health and on the larger ecosystem. Although some concerns can be addressed on a microscale, developing strong documentation to support meaningful activity on a global scale is far more challenging. Prüss-Ustün and coworkers [14] in their review of "knowns and unknowns" of disease burden estimations note that incomplete exposure information limits assessment of the burden of disease from chronic exposure to pesticides and herbicides.

At the time that the Prüss-Ustün report was published, the IARC findings regarding organophosphates and glyphosate had not been issued. In the interim, while technological advances have made it possible to better assess pesticide effects at the cellular level that may lead to the development of cancer, it is difficult to apply such findings to exposed populations. Thus, many of the specific points cited by Prüss-Ustün and coworkers in 2011 continue to apply.

A number of issues regarding the global use of pesticides and herbicides were addressed eloquently in the UN document (cited above) that was released in March 2017. The document, a joint report issued by the UN Special Rapporteur on the right to food and the UN Special Rapporteur on the UN Human Rights Council [20], considered the global right to food and laid out several issues that relate to concerns about pesticide use (a broad category that encompasses herbicide use as well), including the inequitable effects on health resulting from its global use. The report addresses adverse effects on human rights, including human health and environmental health; identifies existing legal structures, nationally and internationally, that could address these questions; provides a summation of challenges (as of 2017); and suggests viable alternatives to the present heavy use of pesticides. It is important to stress that the report examines issues of all pesticide use and therefore would include the use of organophosphate pesticides and glyphosate.

As well as concerns about their use in specific locations, the world-wide movement of pesticides poses additional problems. As noted, pesticides banned in more developed countries may, for numerous reasons, be used in less-developed ones, and a significant portion of their use is tied to agribusiness, a growing global activity. Today, given the nature of agricultural production, the chemical-dependent nature of agribusiness is a challenge that cannot be effectively addressed by prohibitions that exist only in more developed geographic entities. Ironically, many of these same entities import agricultural products from the less-developed locales where pesticide use is not well-regulated. This process potentially poses problems both for agricultural workers in these specific locales but also for those living in countries that import the products.

12.2 Global Production and Movement of Agricultural Produce, Plants, and Flowers

To understand the impact of pesticide and herbicide use on farm workers and those involved in plant production, it is necessary to first consider the extent and growth of global production. The following narratives are not inclusive, but will, hopefully, provide some clarity about trends in production and the global movement of several sectors of agricultural activity.

12.2.1 Production and Global Movement of Agricultural Produce

The term "agricultural produce" encompasses a wide variety of fruits and vegetables, many of which must be grown in defined climates. It is not surprising, therefore, that more developed countries, especially geographic areas with shorter growing seasons, having the resources to import these products do so. It is perhaps also not surprising that the volume of many imports is increasing. A US Department of Agriculture publication [7] following the pattern of imports and exports from the 1970s through 2015 notes that, in the 1970s, the US exported more agricultural products than it imported. However, this pattern began to change in the mid-1990s, and by 2015 the net trade ratio of exports to imports was roughly —10%. While it would be disingenuous to attribute this change only to the growth in demand for imports, the growth, availability, and increasingly widespread development of agribusiness, as well as various price supports all play important roles.

For the US the leading exports are to countries including Mexico, Canada, Chile, the European Union, China, Peru, and Costa Rica. A meaningful number of imports to the US are crops that are more easily grown in tropical countries. Not surprisingly, therefore, bananas, tropical fruits, several other fruits, and coffee account for a significant amount of imports.

Like the US, countries in the EU import more fresh vegetables and fruit than they export. In 2017, approximately 87% of agricultural produce items imported to the EU were varieties of fruit [2]. Interestingly, however, in 2017, for the EU the major source of nuts (considered to be fruit in this classification) was the US. South Africa was a major exporter to EU nations of citrus fruit, cherries, apricots, plums, grapes, apples, and pears.

In Asia, Japan imports a significant amount of its fresh vegetable and fruit supply. A report comparing mid-year 2020 statistics with those of 2019 found a smal net increase, from 873,714 tons (approximately 963,000 US tons) in 2019 to 897,458 tons (approximately 989,000 US tons) in 2020 [8]. China is a major exporter to Japan, although the US is the major exporter of processed foods (canned or frozen foods), including frozen corn, potatoes, and mixed vegetables [5].

Rice, a staple in many diets, is a commodity exported by the US as well as several other countries. In 2018, according to OEC [12], primary exporters of rice were India, Thailand, Vietnam, Pakistan, and the US; top importers were China, Iran, Saudi Arabia, Benin, and Indonesia.

12.2.2 Production and Global Movement of Agricultural Fibers

Fibers also move globally. The two most important of these, cotton and linen (or flax) are produced in several countries. According to data from 2019, in terms of US dollars, the five major cotton exporting countries were China, the US, India, Vietnam, and Brazil [23]. In terms of volume, Asian countries produced the largest amount of cotton, followed by North America, Europe, Latin America, Africa, and Oceania [23]. Interestingly, while in general, countries importing cotton are not the same as those exporting cotton, in 2018–2019, China was reported to both import and export the largest amount of cotton in terms of US dollars [24]. Although it has not produced as much cotton for export in the past four to five years, Pakistan nevertheless remains a major producer of cotton for export [19].

Flax, the plant that produces the flax fibers used to make linen, is not produced in the same volume but has has grown in production over the past few years [11]. According to the Observatory of Economic Complexity, in 2018 the major exporters of flax fibers were France, Belgium-Luxembourg, Lithuania, Egypt, and Belarus. At the same time, the major importers were China, Belgium-Luxembourg, India, France, and Lithuania [11].

12.2.3 Production and Global Movement of Plants and Flowers

The increasing demand for cut flowers, ferns, and decorative plants also contributes to the growth in the global movement of agricultural products. A recent USDA document [21] notes that imports of plant products grew approximately 375% form 2000 to 2018. At the same time, as production is increasing, a global movement has become increasingly diversified in terms of the countries involved. As has been true for many years, The Netherlands continues to remain the greatest producer and exporter of cut flowers and other decorative plants. Nevertheless, a number of other countries have developed floriculture and plant agricultural operations that facilitate global movement, and these countries are increasingly competitive in the world market [13, 16]. In 2021, other countries producing and exporting plants and flowers include (in rank order) Colombia, Ecuador, Kenya, Belgium, Ethiopia, Malaysia, Italy, Germany, and Israel [13]. In 2019, the top five flower exporting countries (in rank order) were The Netherlands, Colombia, Ecuador, Kenya, and Ethiopia [13]. Nevertheless, the movement of plants and flowers is, to a large extent, dependent on geographic locale [16]. Movement in the EU is from EU countries as well as Kenya and Ethiopia; in Asia, it is from countries such as Malaysia and Israel; and in the Northern hemisphere movement to countries including the US and Canada is from Colombia and Ecuador. However, as noted, these movement patterns are not long-standing and may change over time because of other considerations. For example, in 2016, on a global basis, Colombia moved more cut flowers than any other country because, in large part, of the smaller expense of shipping by sea using shipping containers [22]. It is reasonable to expect that, as business practices change, the routes and volume of movement of flowers and plants from one geographic locale to another may also change.

12.3 Global Challenges to Human Health

As the movement of fruits and vegetables, fibers, flowers, and plants become more global, concerns about pesticide and herbicide use in developing countries have also become more pressing. The United Nations Human Rights Council points out that (as noted earlier).

> Lack of harmonized standards.....results in more toxic, and even banned pesticides being used extensively in developing countries because they are cheaper alternatives. In many cases, highly hazardous pesticides that are not or no longer permitted for use in industrialized countries are exported to developing countries. Some pesticide companies fail to register or reregister products intended for export to developing countries, or increase exports of products that have been banned or restricted to use up existing stocks [20].

Clearly, it is true that long-term exposure to numerous pesticides and herbicides used in the global production of farm, flower, and plant crops can lead to the development of chronic disease. While cancer is one potential chronic health condition resulting from exposure, other chronic endpoints are also of concern. Both IARC and WHO have established a number of processes to better follow the development of chronic disease, including the development of networks among countries in specified geographic areas. However, a continuing need for more extensive world-wide data regarding health responses exists.

While many of the issues associated with pesticide and herbicide exposure to workers in developed countries are also those faced by workers in less developed countries, other factors may exacerbate adverse responses. It is useful to consider the other factors that contribute to the challenges involved in pesticide handling and protections that are more widespread in underdeveloped farming communities. This applies to crops that are produced in large-scale farms as well as pesticide residues in products intended for local consumption. Interestingly, the bulk of published reports describe relatively consistent patterns across a number of underdeveloped countries. Not only are the problems consistent among different countries, but it is consistently difficult to pin-point specific exposures. Many pesticides and herbicides are used in combination, raising the potential of synergistic effects. In addition, the marketed forms of many contain adjuvants that may help in their applications to crops but may also be toxic. Among pesticides and herbicides used, malathion and diazinon, as well as glyphosate, are reported in several studies.

Movement of toxic pesticides and herbicides that are restricted in developed to less-developed locales, their use in less developed locales, and the potential for some residues to return to countries where their use has been restricted close to what is described by Faber as "the circle of poison" [4]. This phenomenon has been especially prominent among floral workers. To prevent the movement of pests from one country to another, imported cut flowers have been found to be contaminated with a number of pesticides, including small residues of diazinon [18].

Many issues associated with pesticide use and exposure in less-developed countries may be intensified by conditions in these same locales. While some challenges are not different from those found in developed countries such as the US, others are Greater availability of pesticides banned or restricted in developed countries, poor regulations, additional difficulties for workers, as well as working conditions in many agricultural facilities introduce additional problems. Aspects of these are discussed below.

12.3.1 Availability of Pesticides Banned or Restricted in Developed Countries

As discussed earlier, pesticides banned or restricted in developed countries are often moved to other countries having less strict regulations. In these countries they are often sold with few instructions about how they should properly be used, information about their toxicity or appropriate protective equipment. Potential negative effects are compounded by a number of these factors as discussed below.

12.3.2 Level of Education and Literacy

In many less-developed countries the level of education is very low. The level of education is closely tied to the literacy for many agricultural workers, and in some cases, this challenge is compounded by the number of different indigenous languages (at times not recorded) spoken in some locales.

As noted earlier, literacy among farm workers in developed countries often poses problems, but these problems may be intensified in less-developed constituencies where educational opportunities may be fewer. Numerous examples of such challenges exist. A comparative study of farmers working on small farms in Costa Rica and Uganda found the level of education to be approximately 6 years (SD 2.8) in Costa

Rica and 8 (SD 3.8) among farmers in Uganda [17]. In both countries several indigenous languages are spoken. Chouaibou and coworkers, studying pesticide handling in rice farms of Côte d'Ivoire, found that the bulk (approximately 75%) of farmers had not attended school, and approximately 15% had "been once in secondary school" [1]. While French is the official language, five other languages are widely spoken and it is estimated that approximately 78 languages are indigenous to the country. A study in the Arusha area of Tanzania examining practices of farmers handling pesticides reported that a large majority of farm workers had less than seven years of school [3]. English and Swahili are the official languages, although a number of others are also spoken. A study among agricultural workers on coffee growing facilities, both organic and conventional, in the Dominican Republic, reported a somewhat higher level of education among organic farmers: approximately 33% of organic farmers versus 21% of conventional farmers had completed secondary education [6], although the difference was not significant. A form of Spanish is the official language, although two indigenous languages also are spoken. In many cases, irrespetive of geographic locale, the more frequently spoken indigenous languages are not written.

12.3.3 Procurement and Handling of Pesticides and Herbicides

Proper handling and use of pesticides and herbicides are problems in many locales. Beyond the need for directions to be available and for pesticide users to be literate, handling and use encompass many other issues: procurement, preparation of substances for application, proper application, pesticide storage, and pesticide disposal. Many of these issues are exacerbated by the lack of running water or adequate electricity, or both. The handling practices are well-documented in a publication of Quansah et al. [15]. Examples of these practices in a number of areas are given in the following paragraphs [9].

While procurement should involve obtaining pesticides and herbicides from licensed vendors, a number of reports stated that pesticides were frequently procured from unauthorized vendors [1, 3], which include unlicensed sellers on roadside stands [1]. One study noted that some substances were repackaged into "secondary" containers (implying that these were also sold by an unauthorized vendor and that instructions for their use were not available) [3]. At times, these containers were found to be unlabeled [3].

An additional challenge is understanding the proper use of the pesticides and herbicides, an issue that encompasses several issues: lack of adequate training [1, 10, 17]; not using recommended dose [1]; or pesticide preparation by individuals without adequate guidance [1, 10, 17]. In some cases, substances were mixed "by hand" for application [15, 25].

Pesticide and herbicide storage and disposal were also reported to be problematic. These substances have been found stored in homes, near food, and near bedrooms

[10, 25]. The disposal occurred in a number of ways: "dumping" on farms, in rivers, burying, or selling to pesticide retailers [1, 6]. In some cases, empty containers were left in unsafe locations [10]. Water used for cleaning applicators was reported to be disposed of in rivers, gardens, and soil [6].

The use of personal protective equipment was generally poor. While it is true that the use of personal protection may be difficult in warm climates, it is essential. As is true of the issues cited above, poor education and lack of literacy may play a significant role in farmers' challenges in choices of personal protection. A study in Pakistan looking at the use of personal protective equipment among Pakistani women involved in cotton-picking found that education was positively correlated with the use of protective equipment. Experience picking cotton also played a role [9].

12.3.4 *Growth of Insect Vectors and Vector Control*

In some farming operations, the accumulation of water is a challenge that may promote the use of pesticides, not to remove plant pests, but to combat other insects. This is especially true in rice farming where flooding is a common practice to produce a crop that is a dietary staple for roughly half of the global population. However, insect vectors such as the mosquitos that cause malaria, dengue or the Zika virus often breed in flooded rice paddies. While appropriate pesticide use may be beneficial, overuse is problematic.

A study of rice farmers in Côte d'Ivoire that looked at the use of pesticides and herbicides [1], both in rice farming and in insect vector control, found that pesticides exert a selective pressure for the development of mosquito resistance. A laboratory examination of resistance that examined only malaria-causing mosquitos, found that fewer mosquitos developed resistance when treated with malathion as compared with other pesticides. Nevertheless, a meaningful number did. The authors point out many farmers do not understand the development of pesticide resistance, and, lacking this understanding, may interpret this as the result of "underdosing." Thus, they would increase the amount of pesticide that is used. This phenomenon of was also cited by others.

12.4 Global Data on Chronic Conditions Related to Pesticide and Herbicide Exposure

The development of meaningful data relating pesticide and herbicide exposure to chronic health conditions, including cancer and pre-cancerous conditions, requires both detailed information about exposure and accurate data on disease outcomes. As noted earlier by Prüss-Ustün [14], global exposure information is limited, compromising the ability of epidemiologists and other investigators to accurately determine the magnitude and duration of exposure for agricultural and floriculture workers. Also lacking are data about the development of chronic conditions. While both WHO and IARC have been taking steps to establish networks among countries within established geographical areas to both facilitate better communication, and to develop a depository for exposure and outcome information, more remains to be done. For example, it is unclear what proportion of farm workers who develop clinical conditions either seek or are able to find health care practitioners who can diagnose and treat their conditions. Equally unclear is whether practitioners understand the relationships between pesticide or herbicide exposure and health conditions, given that a number of conditions might result from multiple causes, and that it may be difficult to pin-point specific exposures, since many workers are exposed to several substances.

Several reports regarding pesticide and herbicide use describe symptoms that are relatively non-specific and might be the result of numerous causes. For example, a detailed cross-sectional study conducted by Quansah and coworkers [15] looked at a number of health endpoints that could be related to long-term pesticide exposure. This included self-reported respiratory symptoms such as coughing, wheezing, and breathlessness. Non-respiratory symptoms included depression, sexual impotence, weakness, and an inability to concentrate. The majority of farmers involved in the study had done agricultural work for more than 12 years. Diazinon was among a large number of pesticides to which the farmers had been exposed. A major focus of the study was on health outcomes, but it also focused on pesticide handling practices that were common in underdeveloped countries. The practices were found to correlate with negative health outcomes for the health endpoints considered. However, because this was a cross-sectional study, results can suggest relationships between endpoints and the types of exposure, but cannot unequivocally establish any specific relationship between exposure and outcome.

A study conducted in a coffee-growing area of the Dominican Republic [6] contrasted findings of genotoxicity in buccal cell samples taken from conventional coffee growers who mixed and applied pesticides to buccal cell samples taken from organic coffee growers who had not been occupationally exposed to pesticides for a period of five years or more. While the study was also cross-sectional, and not all demographic characteristics of the farmers in the cohort of conventional coffee growers and those in the organic coffee grower cohort were possible to match, conventional growers were found to handle pesticides with little or no personal protective equipment. All genotoxic markers in buccal cell samples taken from the conventional grower cohort were found to be statistically greater than markers found in the organic farmer cohort. It is noteworthy that one of the pesticides and herbicides used most frequently in conventional coffee growing was glyphosate.

The coffee-growing area had no electricity or running water, a condition not uncommon in farming communities located in underserved countries. To analyze buccal cell samples for markers of genotoxicity, it was necessary to use a protocol that involved sample preservation and transport for further analysis to a laboratory in Vienna, the home institution of the investigators. The authors note that an advantage of the investigation is that it demonstrated the feasibility of collecting genotoxicity information about pesticide-exposed agricultural and nursery workers working in environments where technical facilities are not readily available. Such a model could be applied to similar farms and nurseries in a number of underserved environments and may be a significant way of measuring exposure/outcome relationships in these communities.

12.5 Re-Examining Pesticides and Herbicides: The Importance of IARC and WHO

The results of studies such as the two cross-sectional studies cited above, taken together with observations of how pesticides and herbicides are prepared for application, handled, applied, stored, and disposed of in a number of underserved communities provide strong evidence that the magnitude of health threats posed by pesticides and herbicides may be seriously underreported. Although the purpose of the IARC monographs is to identify and characterize carcinogenic hazards, not to assess risk, the risks assigned to a number of IARC-evaluated pesticides by other agencies, risks that have been assessed in developed countries, may significantly underestimate the danger to a large segment of the global population. This is as true for the four organophosphate pesticides and glyphosate as it is for a number of other widely used substances.

By their nature, pesticides and herbicides, especially when widely used, affect plants and animals and the make-up of the larger ecosystem. The focus of this book is four organophosphate pesticides and glyphosate: how widely they are used; how they affect plants, animals, and the larger ecosystem; why they were evaluated by IARC in 2015; the impact of IARC's findings, published in 2017; as well as environmental justice implications of their continuing and frequently unregulated use, both domestically and globally. It is perhaps not surprising that, as data have accumulated, the use of some has been curtailed, at least in developed countries, through regulatory processes. This does not, however, mean that they are unlikely to be used in other locales, or that, at the same time, other pesticides and herbicides will be developed and introduced in developed and underdeveloped countries to replace them. This work is a primary research and marketing goal of agricultural industries. The overall process has the potential to be a repeating and unending cycle, one often supported by arguments regarding food security as raised by agroindustry representatives. As pointed out in the Report of the Special Rapporteur on the Right to Food [20],

> [pesticide manufacture and use] has certainly helped to keep agricultural production apace of unprecedented jumps in food demand. However, this has come at the expense of human health and the environment. Equally, increased food production has not succeeded in eliminating hunger worldwide. Reliance on hazardous pesticides is a short-term solution that undermines the rights to adequate food and health for present and future generations.

The report further notes that "without or with minimal use of toxic chemicals," larger yields of healthier nutritious food can be produced, and lists aspects of what it designates as "agroecology," an alternative approach to extensive pesticide and herbicide use. In support of this argument, the report discusses a number of issues. Included are the following:

1. The use of organic practices has demonstrated that fewer or no pesticides are possible.
2. The arguments regarding food security are inaccurate or misleading. Lack of adequate food supplies can result from blocks in equitable production and/or distribution that result in poor availability for those needing agricultural produce.
3. The need for a robust farming system in which crops are cultivated in suitable locations. When the diversity of cropland is reduced, ecosystems that have predators for crop pests are more likely to be preserved. Such practices do not enhance biodiversity.
4. Rather than devoting large amounts of cropland to agricultural industry farming with its heavy dependence on pesticides, herbicides, and chemical fertilizers, the practices of crop rotation and planting cover crops, both of which help to reduce soil pathogens and increase the organic content of soils should be used more frequently.

As noted in the introduction to this book, pests have been a problem in crop (and by extension, flower and decorative plant) growth throughout history, and, to some extent, early measures adopted to control pests involved home-grown production of highly toxic substances. However, as more pesticides and herbicides were identified, their production was no longer primarily an individual, but an industrial, for-profit, endeavor. As long as given toxicity criteria (generally acute toxicity) were met, human and ecosystem health became secondary considerations. This stands in contrast with the health-focused activities of WHO and IARC.

Pesticides and herbicides do have a role to play in protecting health, and, at times, addressing plant and pest "invasions" that may be difficult to control with other means. But pesticides and herbicides should be used only when needed and used wisely with a central focus on human, other animal, plant, and ecosystem health.

References

1. Chouaïbou MS, Fodjo BK, Fokou GA, Allassane OF, Koudou BG, David J-P, Antonio-Nkondjio C, Bonfoh B, Ranson H (2016) Influence of the agrochemicals used for rice and vegetable cultivation on insecticide resistance in malaria vectors in southern Côte d'Ivoire. Malar J 15(1):426
2. De Cicco A (2019) The fruit and vegetable sector in the EU - a statistical overview, Eurostat
3. Lekei EE, Ngowi AV, London L (2014) Farmers' knowledge, practices and injuries associated with pesticide exposure in rural farming villages in Tanzania. BMC Public Health 14. DOI: https://doi.org/10.1186/1471-2458-14-389
4. Faber D (1993) Environment under fire: imperialism and the ecological crisis in Central America. New York Monthly Review Press
5. Foreign Agricultural Service (2019) Japan: Japanese processed vegetable market 2019. F. A. Service. Office of Agricultural Affairs Tokyo, USDA Foreign Agricultural Service, 2019. https://www.fas.usda.gov/data/japan-japanese-processed-vegetable-market-2019

6. Hutter H-P, Khan A, Lemmerer K, Wallner P, Kundi M, Moshammer M (2018) Cytotoxic and genotoxic effects of pesticide exposure in male coffee farmworkers of the Jarabacoa Region, Dominican Republic. Int J Environ Res Public Health 15(8) 1641. DOI: https://doi.org/10.3390/ijerph15081641
7. Johnson R (2016) The US trade situation for fruit and vegetable products. Congressional Research Service. Washington, DC. https://sgp.fas.org/crs/misc/RL34468.pdf
8. Komorek C (2020) Japan's fresh fruit imports grow. First six months of 2020 show three per cent increase in imports compared to 2019. FruitNet.com. http://www.fruitnet.com/asiafruit/article/182543/japans-fresh-fruit-imports-grow
9. Memon Q, Wagan S, Chunyu D, Shuangxi X, Jingdong L, Damalas CA (2019) Health problems from pesticide exposure and personal protective measures among women cotton workers in southern Pakistan. Sci Total Environ 685(Oct 1):659–666. DOI: https://doi.org/10.1016/j.scitotenv.2019.05.173
10. Ngowi AV, Maeda D, Wesseling C, Partanen TJ, Sanga MP, Mbise G (2001) Pesticide-handling practices in agriculture in Tanzania: observational data from 27 coffee and cotton farms. Int J Occup Environ Health 7(4):326–332
11. Observatory of Economic Complexity (OEC) (2018) Flax Fibers. OEC 4.0. https://oec.world/en/profile/hs92/flax-fibers
12. Observatory of Economic Complexity (OEC) (2018) Rice. OEC 4.0. https://oec.world/en/profile/hs92/rice#:~:text=Rice%20are%20the%20world's%20151st,%2C%20and%20Indonesia%20(%24994M)
13. Petal Republic Team (2020) Global & US floristry and floriculture industry statistics & trends. Petal Republic, New York. https://www.petalrepublic.com/floristry-and-floriculture-statistics/
14. Prüss-Ustün A, Vickers C, Haefliger P, Bertollini R (2011) Knowns and unknowns on burden of disease due to chemicals: a systematic review. Environ Health 10(9). DOI: https://doi.org/10.1186/1476-069X-10-9
15. Quansah R, Bend J, Armah FA, Bonney F, Aseidu J, Yawson DO, Adu MO, Luginah I, Essumang DK, Abdul-Rahaman A, Cobbina S, Iddi S, Tersigni M, Afful S, Osei-Fosu P, Nketiah-Amponsah E, (2019) Respiratory and non-respiratory symptoms associated with pesticide management practices among farmers in Ghana's most important vegetable hub. Environ Monit Assess 101(716)
16. Rikken M (2010) The European market for fair and sustainable flowers and plants. Michiels C. Brussels, Trade for Development Center. https://www.yumpu.com/en/document/read/3235285/the-european-market-for-fair-and-sustainable-flowers-and-plants
17. Staudacher P, Fuhrimann S, Farnham A, Mora AM, Atuhaire A, Niwagaba C, Stamm C, Eggen RI, Winkler MS (2020) Comparative analysis of pesticide use determinants among smallholder farmers from Costa Rica and Uganda. Environ Health Insights, December 14. https://doi.org/10.1177/1178630220972417
18. Tourmi K, Vleminckx C, van Loco B, Shiffers B (2016) Pesticide residues on three cut flower species and potential exposure of florists in Belgium. Int J Environ Res Public Health 13(10). DOI: https://doi.org/10.3390/ijerph13100943
19. Trend Economy. (2021) Pakistan | Imports and Exports | World | Cotton | Value (US$) and value growth, YoY (%) | 2009–2020. https://trendeconomy.com/data/h2/Pakistan/52
20. United Nations Human Rights Council (2017) Report of the special rapporteur on the right to food. Human Rights Council, UN General Assembly. 24 January
21. USDA (2019) Nearly two-thirds of U.S. agricultural imports consist of horticultural and tropical products. USDA Economic Research Service. https://www.ers.usda.gov/data-products/chart-gallery/gallery/chart-detail/?chartId=58362
22. van Rijswick C (2016) World floriculture map 2016: Equator countries gathering speed, Raboresearch. https://research.rabobank.com/far/en/sectors/regional-food-agri/world_floriculture_map_2016.html
23. Workman D (2019) Cotton exports by country. World's top exports. http://www.worldstopexports.com/cotton-exports-by-country/

24. Workman D (2019) Cotton imports by country World's top exports. http://www.worldstopexports.com/cotton-imports-by-country/
25. Wumbei A, Houbraken M, Spanoghe P (2019) Pesticides use and exposure among yam farmers in the Nanumba traditional area of Ghana. Environ Monit Assess 19(5):307. DOI: https://doi.org/10.1007/s10661-019-7449-5

Index

M. Richmond, *Cancer Hazards: Parathion, Malathion, Diazinon, Tetrachlorvinphos and Glyphosate*, AESS Interdisciplinary Environmental Studies and Sciences Series,
https://doi.org/10.1007/978-3-030-81953-8

The manufacturer's authorised representative in the EU is Springer
Nature Customer Service Centre GmbH, Europaplatz 3, 69115 Heidelberg,
Germany. If you have any concerns regarding our products, please
contact ProductSafety@springernature.com

Printed and bound by CPI Group (UK) Ltd, Croydon, CR0 4YY
28/09/2026
02224092-0010